Topics in
Current Physics

26

Topics in Current Physics   Founded by Helmut K. V. Lotsch

Crystal Cohesion and Conformational Energies

Edited by R. M. Metzger

With Contributions by
R. M. Metzger   F. A. Momany   B. D. Silverman
D. E. Williams

With 55 Figures

Springer-Verlag Berlin Heidelberg New York 1981

Professor Robert M. Metzger, Ph. D.

Dept. of Chemistry, University of Mississippi, University, MS 38677, USA

ISBN-13: 978-3-642-81579-9   e-ISBN-13: 978-3-642-81577-5
DOI: 10.1007/978-3-642-81577-5

Library of Congress Cataloging in Publication Data. Main entry under title: Crystal cohesion and conformational energies. (Topics in current physics ; 26). Bibliography: p. Includes index. 1. Crystallography. 2. Crystal lattices. – 3. Conformational analysis. I. Metzger, R. M. (Robert M.), 1940–. II. Series.
QD921.C758   549   81-5267   AACR2

© by Springer-Verlag Berlin Heidelberg 1981

Softcover reprint of the hardcover 1st edition 1981

2153/3130-543210

Preface

With the advent of X-ray diffraction and crystal structure determination in 1912 researchers in physics and chemistry began investigating the problem of crystal cohesion, i.e., on the question of what holds crystals together. The names of M. Born, E. Madelung, P.P. Ewald, F. Bloch, E.P. Wigner, and J.E. Mayer are, in particular, associated with the pre-1940 work on the cohesion of inorganic lattices. The advent of digital computers brought along great advances in the detailed understanding of ionic crystals, molecular crystals, and metals. The work of P.O. Löwdin and A.I. Kitaigorodosky are seminal in these more recent advances.

This volume is a collection of specialist reports on a subset of the general problem of crystal cohesion. It is intended for researchers and advanced students in solid-state chemistry and physics, and biochemistry. WILLIAMS reports on the molecule-independent empirical parameters for dispersion and repulsion that explain, and can predict, the cohesive energy of neutral organic lattices. MOMANY applies similar procedures to the conformational energy problem and shows how they can be used for the pharmacological problems of polypeptide drug design. METZGER uses quantum-mechanical molecule-dependent atom-in-molecule charges, dipole moments, and polarizabilities to study the cohesion of organic ionic (semiconducting) and partially ionic (metallic) lattices. SILVERMAN emphasizes, with quantum-mechanical dimer calculations, the importance of dispersive interactions for the observed stacking modes in organic metallic lattices.

Alas, this book is not encyclopedic: nothing is said about inorganic metals or small inorganic ionic lattices, and many other distinguished workers in the cohesive energy and conformational energy vineyard are not presented here. This modest volume is a progress report that brings together several people whose work has not been reviewed under one cover; it is hoped that this volume will complement other similar collections.

This is my chance to thank the contributors to this volume for their great diligence and forbearance for the delays in getting the book together, to thank Professor A.N. Bloch (Baltimore) for suggesting the book subject, Professor H.J. Keller (Heidelberg) and Dr. P. Delhaes (Bordeaux) for their hospitality during a sabbatical stay in Europe, and last but not least, to thank Chris and Gianni for their support, love, and understanding.

July, 1981   *Robert M. Metzger*

Contents

List of Contributors

Metzger, Robert M.
    Department of Chemistry, University of Mississippi,
    University, MS 38677, USA

Momany, Frank A.
    Department of Chemistry, Memphis State University,
    Memphis, TN 38152, USA

Silverman, B.D.
    IBM Thomas J. Watson Research Center,
    Yorktown Heights, NY 10598, USA

Williams, Donald E.
    Department of Chemistry, University of Louisville,
    Louisville, KY 40292, USA

1. Introduction

R. M. Metzger

The theoretical study of cohesive energies and conformational energies of organic crystals and biopolymers received its first impetus many years ago in the pioneering work of KITAIGORODSKY [1.1]. He suggested that intermolecular interactions and equilibrium geometries in molecular crystals can be understood fairly well if one assumes a properly parametrized dispersion energy to bind the lattice and an ad hoc overlap repulsion energy that prevents the lattice from collapsing to a single point. The necessary classical atom-in-molecule nonbonded parameters can be fitted to experimental data for a few model compounds and hopefully, can be used for most other molecular crystals.

Given the high cost of reliable ab initio quantum-mechanical calculations for large (>50 electrons) systems, Kitaigorodsky's approach still retains its validity today, and its use has indeed brought many interesting developments. This volume of contributed articles attempts to assess the validity of Kitaigorodsky's *Ansatz*. In particular it reviews the systems for which such parametrized calculations have been most successful (neutral molecules, peptides) and analyzes the difficulties encountered (organic ionic and partially ionic crystals). This volume dovetails to some extent with the work described in recent reviews on conformational analysis [1.2] and provides a fresh look at the problem of crystalline cohesion last reviewed seventeen years ago [1.3].

In Chapter 2 WILLIAMS discusses his adaptation of Kitaigorodsky's methods. Non-bonded potential parameters for dispersion and repulsion energies can be obtained from a few model structures and transferred, in general, to other neutral crystals. This procedure yields quantitative estimates of crystal cohesive energies, lattice energy minima, and of lattice intermolecular vibration spectra. In some cases (benzene, chlorine) Madelung potentials and intermolecular Morse potentials become necessary. A review of the Ewald accelerated convergence formulas for the Madelung and dispersion energies is given, along with the necessary expressions for lattice energy minimization with respect to molecular translation and rotation.

Chapter 3 is an account by MOMANY of the application of Momany and Scheraga's computer program for the conformational energy of peptides to current research projects in drug design. Here Kitaigorodsky's *Ansatz* can be used with great success to predict the minimum energy conformer of any peptide. When properly correlated with experimental biological-activity data for closely related peptides, this program can assist in experimental drug design problems by predicting which target peptide, if synthesized, will have the correct conformation for the desired biological activity. Seven case studies are presented to verify the great practical power of this method.

The last two chapters study the difficulties encountered when one uses Kitai-gorodsky-type atom-in-molecule semiclassical calculations to understand the cohesion in organic ionic and partially ionic crystals.

In Chapter 4 METZGER shows that the Madelung energy is, as expected, the preponderant term of the cohesive energy for organic ionic insulators and semiconductors. Here, for the Madelung energy, the atom-in-molecule partial charges are obtained (as in Momany's program) from quantum-chemical calculations for the isolated molecule. However, for the partially ionic organic crystals and particularly for the quasi-one-dimensional organic metals, the Madelung energy and the uniform partially charged lattice model are both found to be insufficient to account for the cohesive energy. The dispersion and repulsion energies are important, but other terms (charge-dipole and polarization) become very important. A careful treatment of the cost of fractional ionization, or else the Wigner crystal limit are needed to explain incomplete charge transfer for these crystals.

Finally, in Chapter 5 SILVERMAN reports on calculations of the stability of dimers of organic electron donors and acceptors and their ions as a function of intermolecular overlap. Minimum-basis semiempirical molecular orbital dimer calculations fail to stabilize the observed intermolecular overlap as it is seen in quasi-one-dimensional metals. On the other hand, the Gordon-Kim density functional algorithm is explicitly designed to account for dispersion interactions of small molecules and gives the correct minima for these dimers. Thus intermolecular dispersion interactions must be crucial in explaining the crystal geometry of such systems.

All in all, the atom-in-molecule semiclassical techniques for crystal cohesion presented here show current validity and vitality, but must be used with care, as must all semiempirical methods. It is hoped that this series of reviews will help and stimulate further research.

References

1.1 A.I. Kitaigorodsky: *Molecular Crystals and Molecules* (Academic Press, New York 1973)
1.2 B. Pullman (ed.): *Quantum Mechanics of Molecular Conformations* (Wiley-Interscience, New York 1976)
1.3 M.P. Tosi: *Cohesion of Ionic Solids in the Born Model*, ed. by F. Seitz, D. Turnbull, Solid State Physics — Advances in Research and Applications, Vol.16 (Academic, New York 1964)

2. Transferable Empirical Nonbonded Potential Functions

D. E. Williams

With 2 Figures

Empirical nonbonded atom parameters can be developed for dispersion, repulsion, and Madelung's (Coulombic) lattice potential energies of a few reference organic crystals, whose crystal structures and experimental lattice energies are known. A review is given of the relevant mathematical procedures (Ewald's accelerated convergence of lattice sums, force-fit and direct-parameter-fit methods).

These nonbonded potential parameters can then be transferred to other crystals and used to predict their lattice energies and lattice vibrations. Progress is also being made in accounting for solid-state molecular and intramolecular rotations, molecular translations, in dealing with structures that may have weak intermolecular bonding (Cl_2), and in predicting the correct pressure polymorph of benzene.

2.1 Introductory Comments

It is natural and practical to express the potential energy of intermolecular interaction in terms of atomic contributions [1.1]. The most abundant and detailed sources of calibration data are molecular crystal structures [1.2]. These crystal structures are normally determined experimentally by X-ray or neutron diffraction. In most cases the goal of the crystal structure determination is to obtain only the molecular structure. In order to solve the crystal structure, it is incidentally necessary to determine the positions of the molecules in the unit cell. These thousands of known crystal structures thus provide an extensive data base for the study of intermolecular potentials. In each crystal the positions of the molecules are dictated by the nonbonded interactions which are present.

A second source of information about nonbonded potential energy is molecular conformation. It appears that the nonbonded interactions between sufficiently distant parts of the same molecule can be described by the same potential functions that describe intermolecular interactions in crystals. Again, there are available many flexible molecular structures which can be used to derive information about nonbonded interactions. This area of interest is referred to as conformational analysis.

Since by definition the nonbonded interactions in conformational analysis are intramolecular in nature, considerable care must be taken to separate bonding and nonbonding effects. A complete conformational potential energy will include bond stretching, bond bending, and internal rotation functions. There are strong correlations between, for instance, bond bending and 1,3 nonbonded interactions. Even 1,4 nonbonded interactions show high correlation with empirical torsion potentials for internal rotations. There is little doubt, however, that the nonbonded forces between distant parts of a large molecule are essentially the same as those between different molecules. Several detailed sets of conformational analysis potential functions are available [2.3].

A third source of information about nonbonded potential energy is theoretical calculations. Recently, dramatic advances have been made in quantum-mechanical calculations of nonbonded interactions [2.4]. Consider the example of the simple molecule, ethylene [2.17]. An extended quantum-mechanical calculation showed that the empirical atom-atom model is a fairly accurate representation of the nonbonded energy of the dimer. However, the atom-atom model potential parameters appear to be fully optimized; no further improvement can be expected without modification of the model. The quantum-mechanical calculations are no doubt capable of much higher accuracy when extended. Further, although the fast evaluation of the atom-atom empirical model is an advantage at present, the availability of extremely fast computers may greatly reduce this advantage in the future.

Macromolecules play crucial roles in living systems. In proteins, for instance, the biological activity depends critically on the folding of the peptide chain to obtain the proper molecular conformation [2.5]. This chain folding is determined largely by nonbonded forces. A recent review is available for nonbonded interactions in macromolecules [2.6].

Many examples of weak bonding are known to be intermediate in energy between true bonds and nonbonds. The most prominent and important case is the hydrogen bond. Obviously, hydrogen-bonded crystal structures can be used as basis data from which to derive an empirical hydrogen bond potential. However, even though each hydrogen bond may be much stronger in energy than a nonbond, there are many more nonbonds than hydrogen bonds in the typical crystal structure. Thus it is usually necessary to have an adequate model for the nonbonded interactions before the model can be extended to hydrogen bonds. Another way of stating this is to point out that although crystals (and macromolecules) exist with only nonbonded interactions, no crystal (or macromolecule) is known which is entirely hydrogen bonded and with nonbonded energy completely absent. In carboxylic acid crystal structures, for instance, the nonbonded energy is greater than the hydrogen bonded energy in most cases. In proteins, also, both the nonbonded and hydrogen-bonded energies must be taken into consideration to correctly predict the conformation.

2.2 Empirical Nonbonded Potential Models

Accurate quantum-mechanical calculations for simple molecules serve as a guide to the selection of practical empirical potential models. The properly calibrated non-bonded potential functions can then be used for large molecules which are presently beyond the reach of accurate quantum-mechanical calculations. The strategy described here is to primarily use quantum-mechanical theory to obtain the functional form of the nonbonded potential energy. Empirical values for the coefficients of the chosen functions are then obtained by fitting experimental data, particularly crystal struc-tures.

Quantum-mechanical theory indicates that the leading and most important term in the nonbonded attractive energy between neutral molecules is proportional to the inverse sixth power of the intermolecular distance [2.7,8]. This energy is referred to as the dispersion energy, V_d. For very simple systems such as the noble gases, higher-order terms in the dispersion energy are often included [2.9]. For poly-atomic molecules the dispersion energy is usually expressed as an atom-atom sum over the nonbonded interatomic distances, r_{jk}, using only the leading term. Some of the error caused by neglect of the higher-order dispersion energy terms is removed by empirical adjustment of the coefficient, A:

$$V_d \simeq -Ar_{jk}^{-6} \ .$$

(2.1)

The dispersion energy has a relatively long range, so that care needs to be taken to include a sufficient number of terms in a crystal-lattice summation to achieve accuracy in the value of the sum. A method of drastically improving the rate of convergence of the lattice sum will be discussed later.

Both quantum-mechanical numerical results for simple systems [2.7,8] and direct experiments [2.10] indicate that the repulsive nonbonded energy can be represented by an exponential function:

$$V_r \simeq B \ \exp(-Cr_{jk}) \ .$$

(2.2)

It is highly desirable to set values of the parameter C so as to reduce V_r to a single-parameter function. (Note that the function

$$V_r \simeq Br_{jk}^{-n}$$

(2.3)

also contains two parameters; there is no reason to arbitrarily set n equal to 12, as is sometimes done.) Some techniques for establishing values of C will be discussed later.

If net charges are present in the molecule, a Madelung (Coulombic) energy term must be considered:

$$V_c \cong q_j q_k r_{jk}^{-1} \quad . \tag{2.4}$$

The approximation sign is still included because there may be some doubt about the location and magnitude of the net charges, even though Coulomb's law is exact. Note that at the atomic level it is not appropriate to include a dielectric constant. The net charges (monopoles) are usually placed on the atoms of the molecule. This procedure is better than the use of point dipoles, point quadrupoles, etc., because the intermultipole distance may be nearly as small as the molecular diameter. The multipole expansion does not converge very well when the molecular charge distribution is nearly as large in diameter as the intermolecular distances. Some techniques for establishing values for the net atomic charges will be discussed later. The Madelung (Coulombic) energy is very long range and the lattice sum converges extremely slowly. Fortunately, a method of drastically improving the rate of convergence is available and this method will also be discussed later.

If weak intermolecular bonding is present, a Morse potential may be added for the particular types of atoms which are interacting:

$$V_m = p^2 D\left\{\exp[-2\beta(r_e-r)]-2\exp[-\beta(r_e-r)]\right\} \quad . \tag{2.5}$$

This empirical bonding function has four adjustable parameters p, D, r_e, and β. The potential has a minimum at r-r_e with energy $V_m = -D$, if p = 1. Sometimes the values of D, r_e, and β are taken from a diatomic molecule, such as molecular chlorine, for use in the treatment of weak intermolecular bonds in the crystal. The single parameter p is then used to express the amount of partial bonding. Note that the numerical value of p does not necessarily have quantitative significance as an indicator of the amount of partial bonding. This situation occurs because the correct values of D, r_e, and β are likely to be different from the free-molecule values.

The complete pair potential energy is obtained by adding together the different contributions:

$$V_{jk} = V_d + V_r + V_c + V_m \quad . \tag{2.6}$$

2.2.1 The Number of Adjustable Parameters and Heteroatomic Combining Laws

Suppose we are dealing with a crystal containing C, H, N, and O atoms whose net charges were already established, and $V_m = 0$. In addition to the presumed known q values, we need to know A, B, and C for ten types of nonbonded interactions. The situation becomes worse, in terms of its requirements for adjustable parameters, if all atoms of a given type cannot be grouped together. In general, the grouping together of atoms will sacrifice accuracy, because we fully expect small differences between A, B, and C even for the same type of atom. In the case of the net charges, the differences can be large.

As an example, consider possible differences between A and B for saturated versus aromatic carbon in hydrocarbons [2.11]. Values of A and B were derived separately from a group of nine saturated hydrocarbon crystals and from a group of nine aromatic hydrocarbon crystals [kJ/mol,Å]:

parameter	saturated	aromatic	combined
A_{CC}	2364	2254	2145
B_{CC}	316000	312200	300300

It is seen that the differences between aromatic and saturated carbon nonbonded potential parameters are small and that grouping these two types of carbon together will cause only minor errors. The decrease in both A and B for the combined data is caused by a better separation between carbon and hydrogen potentials in this case. The better definition of the potentials in the combined data causes an increase in A and B for hydrogen accompanying the decrease in A and B for carbon.

Even if all atoms of a given type can be grouped together, we are left with the need to establish the values of thirty empirical nonbonded parameters in our example. Theory is unable to give quantitative values in most cases at the present time. We can use theory, however, to suggest heteroatomic combining laws for the nonbonded parameters. In the case of V_c, Coulomb's law requires that the geometric mean combining law hold.

According to the classic formula of LONDON [2.12], the dispersion energy depends on the atomic polarizabilities P and ionization energies I:

$$V = -\frac{3}{2}\left(\frac{I_j I_k}{I_j + I_k}\right)\frac{P_j P_k}{r_{jk}^6} \quad . \tag{2.7}$$

Since the range of variation of the polarizabilities is much greater than that of the ionization energies, the geometric mean law is expected to hold fairly well [2.13]. This has been experimentally verified in hydrocarbon crystals [2.14,15]. Thus we can define atomic quantities, a, from which the pair parameters may be formed:

$$a_j = \sqrt{A_{jj}} \tag{2.8}$$

and

$$A_{jk} = a_j a_k \quad . \tag{2.9}$$

For the B parameter, theory is not so immediately helpful. We can postulate that the geometric mean law holds and test the hypothesis with observed crystal structures, e.g., hydrocarbons. Parallel derivations of B were made with and without the assumption of the geometric mean combining law. All net atomic charges were set to zero in this case.

assumption	B_{CH}
no geometric mean	36680
with geometric mean	42090

Thus it appeared that the geometric mean value was too high. Later work [2.16] established that this was an artifact caused by neglect of the net atomic charges in hydrocarbons. This neglect would remove a positive homoatomic contribution to V_c, and a negative heteroatomic contribution. The neglect of the Coulombic heteroattraction causes B_{CH} to decrease in order to compensate.

A test of the necessity for independent variation of B_{CH} can be made, based on the 18 hydrocarbon structures mentioned above. If an independent B_{CH} is needed, the goodness of fit should improve (decrease):

assumptions	goodness of fit
$q_j = 0;\ B_{CH} \neq \sqrt{B_{CC}B_{HH}}$	4.54
$q_j = 0;\ B_{CH} = \sqrt{B_{CC}B_{HH}}$	5.21
$q_j \neq 0;\ B_{CH} = \sqrt{B_{CC}B_{HH}}$	3.26

When B_{CH} was allowed to vary independently in addition to allowing $q_j \neq 0$, no further improvement resulted. The above figures for goodness of fit are completely compatible with the idea that the apparent failure of the geometric-mean law was an artifact of the neglect of net atomic charges. A considerable simplification is achieved; we will assume the geometric mean law for both V_r and V_d (as well as V_c) in further discussions.

For the parameters C, since they are exponents, an arithmetic average is expected. The C values are discussed below. In our example containing C, H, N, and O atoms, we now expect only 4 sets of A, B, C need to be found. All others are set by the geometric mean law.

2.2.2 The Atomic Hardness Parameters, C

When an attempt is made to derive values for C from experimental data, it is realized that there is a very strong correlation between C and B. Thus, if C is set in some manner, B will adjust so as to offset any error in C. If it is intended to optimize the B's, somewhat inaccurate values of C may be adequate.

Nevertheless we seek the best values for C. There are three methods which are being used: (a) direct derivation from crystal structures, especially with use of compressibility or elastic constant data; (b) atomic and molecular beam scattering data; and (c) quantum-mechanical calculations.

Graphite provides a noteworthy case for the direct determination of C for carbon. The carbons in graphite are aromatic; but the value of C obtained should be appropriate for saturated carbon as well, as discussed above. CROWELL [2.18] used the interplanar spacing in graphite, the compressibility, and the lattice energy to obtain values of A, B, and C. We have used Crowell's value of 3.60 Å^{-1} for carbon; the credibility of this procedure is increased by noting that Crowell's values for A_{CC} and B_{CC} are fairly close to those derived directly from hydrocarbon crystals.

KUAN et al. [2.19] derived a value of $C_{NN} = 3.64 \text{ Å}^{-1}$ from the observed lattice energy and lattice frequencies of solid α-nitrogen. They also varied the N-N bond length and obtained 0.9012 Å, as compared to the observed bond length of 1.0976 Å. (See further discussion of bond length foreshortening below.) Calculations were made in the space group Pa3.

LIN [2.20] performed calculations in space group $P2_13$ without bond length foreshortening. He used the observed crystal compressibility rather than the lattice frequencies, along with the crystal structure and energy. His value of $C_{NN} = 3.60 \text{ Å}^{-1}$ is close to that of KUAN et al.

RINALDI and PAWLEY [2.21] fitted the crystal structure, energy, and lattice frequencies of sulfur, α-S_8. An optimum value of $C_{SS} = 2.90 \text{ Å}^{-1}$ was found.

Values for C of the noble gases are also available from crystal structure data. A simplified treatment such as is being discussed here yielded 4.36, 3.58, 3.57, and 3.11 Å^{-1} for Ne, Ar, Kr, and Xe, respectively [2.22]. Very elaborate potentials are available for the noble gases; see [2.9], for example.

Molecular beam scattering experiments provide a different source of information about atomic hardness [2.23]. The region of interest here is that of intermediate energy. The high-energy molecular beam results usually refer to shorter distances than those found in crystals and molecular complexes. The analysis of low-energy scattering is complicated by the necessity for consideration of the weak dispersion attraction as well as the repulsion. Unfortunately, there are serious experimental difficulties with this method in the intermediate-energy range. Except for the noble gases, much of the work involves ionic species (see [2.24], for example). The He_2 dimer has been extensively studied [2.10]. A value of $C_{HeHe} = 3.95$ was found.

Until recently, theoretical quantum-mechanical calculations have been limited to smaller molecules. This method is being extended to larger molecules as fast computers and computer programs become available [2.17]. The $(H_2)_2$ dimer has been extensively studied. Of course, this is the simplest case with which to test the usefulness of an atom-atom intermolecular repulsion model. It should be noted immediately, however, that hydrogen is different from such atoms as C, N, and O in that it does not have a filled inner electron shell, that is, all of the electrons in H_2 are bonding electrons. The atoms in H_2 may therefore be less spherical than atoms with a higher atomic number. The atomic hardness of hydrogen is discussed below in connection with the concept of a repulsion center shift or bond foreshortening.

2.2.3 The Repulsion Center Shift for Bonded Hydrogen

For a number of years, crystallographers were puzzled by the observation of shorter C-H bond lengths measured by X-ray diffraction, as compared with neutron diffraction or microwave spectroscopy. This discrepancy was explained when it was realized that the X-ray results were indicating a shift of electron density into the bond, while the other methods indicated the nuclear positions. This apparent bond foreshortening is expected to be most important for hydrogen, where no inner shell electrons are present.

STEWART et al. [2.25] fitted two spherical density functions to an accurate electron density of H_2, allowing the spherical centers to float away from the nuclear positions. The best fit was obtained when the electron density centers were shifted 0.07 Å into the bond. Since the repulsion energy between molecules with filled orbitals is approximately proportional to the electron-density overlap, the repulsion centers are also shifted into the bond by an equal amount.

Additional evidence for the repulsion center shift comes from quantum-mechanical calculations for $(H_2)_2$. This calculation has been carried out for several intermolecular distances and for four different orientations in the dimer. These orientations are linear (end to end); square planar (side to side); T-shaped (end to side); and tetrahedral (side to side, rotated 90°) [2.26]. The intermolecular energy was fitted to an atom-atom model (dumbbell model) with good success, provided the repulsion centers were allowed to shift into the bond [2.27]. The fitting process yielded empirical values for B_{HH}, C_{HH}, and the bond foreshortening: 4357 kJ/mol, 3.14 $Å^{-1}$, and 0.16 Å, respectively.

Thus, three different approaches verify that a repulsion center shift occurs in a bonded hydrogen atom. Pending a more accurate value, the 0.07 Å shift found by STEWART et al. may be adopted as a conservative estimate of the C-H bond foreshortening in hydrocarbons. For accurate work it is important that the same foreshortening be used in applications of the nonbonded potentials as was used in the derivation of the potentials.

2.2.4 The Net Atomic Charges, q

For restricted classes of molecules (e.g., hydrocarbons or perchlorohydrocarbons) there may be only one or only a very small number of independent net atomic charges, and these may be determined from crystal structure data. For more general types of molecules there are too many different net atomic charges to determine.

It is often the case that the Madelung (Coulombic) component of the lattice energy is only a small fraction of the total lattice energy. If the Coulombic component is found by subtraction of the (exp-6) empirical energy from the observed lattice energy, any errors in the (exp-6) portion could be greatly magnified in the difference found for the Coulombic component. Therefore, it is desirable to have a more direct method of determining the net atomic charges if they are small.

In the case of many hydrocarbon crystals the complete neglect of net atomic charges has only a minor, but nevertheless significant, structural effect (see further discussion in 2.5.2). However, BATES and BUSING [2.37] showed that if net atomic charges were neglected in hexachlorobenzene the predicted molecular orientation angle in the crystal was off by $14°$. They found that if a net atomic charge of -0.1058e was placed on the chlorines, with corresponding positive charges on the carbons, the molecular orientation was reproduced within $2°$.

The net atomic charge of an atom in a molecule is not an exactly defined physical quantity in quantum mechanics. However, the electrostatic potential at points in space surrounding the molecule is an exactly defined property. SCROCCO and TOMASI [2.66] review calculations of the molecular electrostatic potential using the self-consistent field molecular orbital (SCF-MO) method. MOMANY [2.67] determined net atomic charges by empirically fitting them to the SCF-MO calculated electrostatic potentials of formamide, methanol, and formic acid. COX and WILLIAMS [2.68] investigated the method further by obtaining potential-derived charges (PD charges) for hydrogen fluoride, water, ammonia, methane, acetylene, ethylene, carbon dioxide, formaldehyde, methanol, formamide, formic acid, acetonitrile, diborane, and carbonate ion. They used the Gaussian basis sets STO-3G, 6-31G, and 6-31G** to obtain the molecular wavefunctions.

The PD-charge method neglects any polarization of the molecule in going from the gas to the crystal. There is some evidence that polarization is not a very large component of the energy of weak intermolecular complexes. Even in the case of hydrogen bonded complexes the energy decomposition studies of UMEYAMA and MOROKUMA [2.69] showed that polarization effects are not too large. Therefore, the PD-charge method appears promising for weaker complexes.

SMIT, DERISSEN, and VAN DUIJNEVELDT [2.70] have used this method of determining net atomic charges in methanol, formaldehyde, and formic acid. They applied their results to the structures of formic acid and acetic acid dimers, and to the crystal structures of formic acid, acetic acid, α-oxalic acid, and β-oxalic acid. COX, HSU, and WILLIAMS [2.71] have also recently used the PD-method for estimation of the net atomic charges in the crystal structures of trioxane, tetroxocane, pentoxecane, succinic anhydride, diglycollic anhydride, p-benzoquinone, furan, and 1,4-cyclohexanedione.

2.3 Nonbonded Interactions in Molecular Crystals

A very large number of molecular crystal structures have been determined by X-ray diffraction; some additional structures have been determined by neutron diffraction. There are over 20,000 crystal structure determinations listed in the Cambridge organic-crystal-structure data file [2.2]. The molecular weight ranges from substances that are liquid at room temperature to complicated proteins and polynucleotides. Most structures are determined at room temperature, although structure determinations down to liquid nitrogen temperature are fairly routine, and a few structures have been investigated at liquid helium temperature. Although in most cases the experimental objective was to determine the molecular structure, the intermolecular packing structure must also be accurately determined in each case. A wealth of information about nonbonded interactions is contained in these packing structures.

2.3.1 Nonbonded Crystal Lattice Energy

The prototype molecular crystal for our discussion has one rigid molecule with N atoms in the asymmetric part of the unit cell. The remaining molecules can be found from the unit cell constants and the space group. The nonbonded energy per mole is obtained by a pairwise summation over the distances r from each atom in the asymmetric unit to the atoms in all surrounding molecules.

$$V = \frac{1}{2} \sum_{\substack{j=1 \\ \text{asymmetric} \\ \text{unit}}}^{N} \sum_{\substack{k \\ \text{surrounding} \\ \text{molecules}}}^{\infty} V_{jk}(r) \quad . \tag{2.10}$$

Our prototype molecular crystal has twelve structural degrees of freedom: six unit cell constants $(a,b,c,\alpha,\beta,\gamma)$, three rotations of the molecule $(\theta_1,\theta_2,\theta_3)$, and three translations of the molecule (X_1,X_2,X_3).

The observed space group may place restrictions on the allowed degrees of freedom. For instance, in the monoclinic system α and γ are fixed at $90°$; or in the tetragonal system b must be equal to a. In polar space groups there are fewer than three translations allowed, e.g., in P2 only X_1 and X_3 can be varied.

If there is more than one molecule in the asymmetric unit, an additional set of three rotations and three translations are added for each additional molecule. If there is less than one molecule in the asymmetric unit, there will be restrictions on the rotations and/or translations of the molecule.

Consider the fairly common case of a molecule situated on an inversion center. This situation is conveniently handled by redefining an apparent asymmetric unit composed of the entire molecule (with an inversion center present in the coordinate list), and by holding X_1, X_2, and X_3 invariant. Similarly, if the molecule has two-fold symmetry along b, the entire molecular coordinate list is used, but θ_1, θ_3, X_1, and X_3 are not varied.

Let us label the crystal structure's degrees of freedom with the parameters p_i and the nonbonded potential function parameters as q_i. The potential energy of the crystal depends on both $\underline{p}$ and $\underline{q}$. For a given $\underline{q}$ and at constant entropy and volume, the stable state is at a minimum in V. Thus the gradient of V is zero at equilibrium:

$$\underline{F}_{-p} = \frac{\partial V(\underline{p}^0, q)}{\partial \underline{p}} = 0 \quad . \tag{2.11}$$

In other words, the forces and torques are zero at equilibrium.

Since the entropy of the crystal at constant temperature only varies slowly with the crystal structure, the neglect of ΔS is not very serious. Also, the observed thermal expansion of crystals is generally small, with only a slow variation of $\underline{\theta}$ and $\underline{X}$ with temperature. Further discussion of thermal effects will appear later. Since the thermal expansion is small, pressure-volume work is negligible at atmospheric pressure for crystals; only at high pressures does the pressure-volume work of unit cell expansion become significant.

2.3.2 Mathematical Description of the Crystal Structure

Let $\underline{X}_j$ be the Cartesian coordinates [Å] of the atoms in the apparent asymmetric unit in the zeroth (origin) unit cell, and let $\underline{X}_{km}$ be the atomic coordinates for the surrounding molecules. The nonbonded distances are

$$\underline{r}_{jkm} = |\underline{X}_j - \underline{X}_{km}| = \underline{C} \quad , \tag{2.12}$$

where the subscript m designates a symmetry operation.

In unit-cell space the symmetry operations are conveniently specified by 3×3 matrices $\underline{s}_m$ and by vectors $\underline{t}_m$ operating on the fractional cell coordinates x_k:

$$\underline{x}_{km} = \underline{s}_m \underline{x}_k + \underline{t}_m \quad . \tag{2.13}$$

The $\underline{t}_m$ include cell translations as well as those due to screw axes and glide planes.

Crystal unit-cell space may be transformed to reference Cartesian space by the matrix $\underline{D}$. If we choose Cartesian axes $\underline{e}_y$ and $\underline{e}_z$ coincident with $\underline{b}$ and the reciprocal axis $\underline{c}^*$ respectively, and $\underline{e}_x$ in the $\underline{ab}$ plane perpendicular to $\underline{bc}^*$, the elements of $\underline{D}$ are

$$\underline{D} = \begin{bmatrix} a \sin\gamma & 0 & c(\cos\beta - \cos\alpha \cos\gamma)/\sin\gamma \\ a \cos\gamma & b & c \cos\alpha \\ 0 & 0 & V/(ab \sin\gamma) \end{bmatrix} , \tag{2.14}$$

where

$$V = abc(1 - \cos^2\alpha - \cos^2\beta - \cos^2\gamma + 2 \cos\alpha \cos\beta \cos\gamma)^{\frac{1}{2}}$$
$$\equiv abc\delta \tag{2.15}$$

Using capital letters for the operators in Cartesian space, we have Cartesian atom positions $\underline{X}_{km}$ given by

$$\underline{X}_{km} = \underline{S}_m \underline{X}_k + \underline{T}_m \quad , \tag{2.16}$$

where

$$\underline{S}_m = \underline{D}\underline{s}_m\underline{D}^{-1} \quad \text{and} \quad \underline{T}_m = \underline{D}\underline{t}_m \quad . \tag{2.17}$$

The Cartesian components of T are

$$
\begin{aligned}
T_1 &= t_1 a \sin\gamma + t_3 c(\cos\beta - \cos\alpha \cos\gamma)/\sin\gamma \quad , \\
T_2 &= t_1 a \cos\gamma + t_2 b + t_3 c \cos\alpha \quad , \\
T_3 &= t_3 c\delta/\sin\gamma \quad .
\end{aligned}
\tag{2.18}
$$

Note that a change in the lattice constants affects only $\underline{T}$ and leaves the $\underline{X}_j$ invariant. The Cartesian coordinates of the molecular center may be affected, but we assume that the fractional coordinates of the center are invariant. This assumption is convenient, for example, when a centrosymmetric molecule is required to retain its nonorigin position in the cell as the cell constants vary. The resulting molecular Cartesian translation is

$$\Delta\underline{X} = \underline{X}^0_{new} - \underline{X}^0_{old} = \underline{D}\underline{x}^0 - \underline{X}^0_{old} \quad . \tag{2.19}$$

The expression for $\underline{r}_{jkm}$ becomes

$$\underline{r}_{jkm} = \underline{C} = (\underline{X}_j + \Delta\underline{X}) - \underline{S}(\underline{X}_k + \Delta\underline{X}) - \underline{T} \quad . \tag{2.20}$$

We now can prepare expanded equations for the components of $\underline{C}$ and $\underline{X}$:

$$
\begin{bmatrix} C_1 \\ C_2 \\ C_3 \end{bmatrix}
=
\begin{bmatrix} X_{1j}+\Delta X_{1j} \\ X_{2j}+\Delta X_{2j} \\ X_{3j}+\Delta X_{3j} \end{bmatrix}
-
\begin{bmatrix} S_{11} & S_{12} & S_{13} \\ S_{21} & S_{22} & S_{23} \\ S_{31} & S_{32} & S_{33} \end{bmatrix}
\begin{bmatrix} X_{1k}+\Delta X_{1k} \\ X_{2k}+\Delta X_{2k} \\ X_{3k}+\Delta X_{3k} \end{bmatrix}
-
\begin{bmatrix} T_1 \\ T_2 \\ T_3 \end{bmatrix}
\tag{2.21}
$$

$$
\begin{bmatrix} \Delta X_1 \\ \Delta X_2 \\ \Delta X_3 \end{bmatrix}
=
\begin{bmatrix} a \sin\gamma & 0 & c(\cos\beta - \cos\alpha \cos\gamma)/\sin\gamma \\ a \cos\gamma & b & c \cos\alpha \\ 0 & 0 & c \delta/\sin\gamma \end{bmatrix}
\begin{bmatrix} x_1^0 \\ x_2^0 \\ x_3^0 \end{bmatrix}
-
\begin{bmatrix} x_1^0 \\ x_2^0 \\ x_3^0 \end{bmatrix}
\tag{2.22}
$$

$$
=
\begin{bmatrix} ax_1^0 \sin\gamma + cx_3^0(\cos\beta - \cos\alpha \cos\gamma)/\sin\gamma - x_1^0 \\ ax_1^0 \cos\gamma + bx_2^0 + cx_3^0 \cos\alpha - x_2^0 \\ cx_3^0\delta/\sin\gamma - x_3^0 \end{bmatrix}
\quad .
\tag{2.23}
$$

The final expressions for C_1, C_2, and C_3 are given below. (We use subscripts j and k on $\underline{x}^0$ to allow the possibility that there is more than one molecule in the asymmetric unit. In that case, $\underline{x}^0$ will have several values, each applying to a certain range of j or k.)

$$C_1 = -(t_1 - x_{1j}^0 + S_{11} x_{1k}^0) a \sin\gamma - (t_3 - x_{3j}^0 + S_{11} x_{3k}^0) c (\cos\beta - \cos\alpha \cos\gamma)/\sin\gamma$$

$$- S_{12} a x_{1k}^0 \cos\gamma - S_{12} b x_{2k}^0 - S_{12} c x_{3k}^0 \cos\gamma - S_{13} c x_{3k}^0 \delta/\sin\gamma + x_{1j}^0 - x_{1j}$$

$$- S_{11} x_{1k} - S_{12} x_{2k} - S_{13} x_{3k}^0 + S_{11} x_{1k}^0 + S_{12} x_{2k}^0 + S_{13} x_{3k}^0 \qquad (2.24)$$

$$C_2 = -(t_1 - x_{1j}^0 + S_{22} x_{1k}^0) a \cos\gamma - (t_2 - x_{2j}^0 + S_{22} x_{2k}^0) b - (t_3 - x_{3j}^0 + S_{22} x_{3k}^0) c \cos\alpha$$

$$- S_{21} x_{1k}^0 a \sin\gamma - S_{21} x_{3k}^0 c (\cos\beta - \cos\alpha \cos\gamma)/\sin\gamma - S_{23} x_{3k}^0 c\delta/\sin\gamma$$

$$+ x_{2j} - x_{2j}^0 - S_{21} x_{1k} + S_{21} x_{1k}^0 - S_{22} x_{2k}^0 + S_{22} x_{2k}^0 - S_{23} x_{3k}^0 + S_{23} x_{3k}^0 \qquad (2.25)$$

$$C_3 = -(t_3 - x_{3j}^0 + S_{33} x_{3k}^0) c\delta/\sin\gamma - S_{11} x_{1k}^0 a \sin\gamma - S_{31} x_{3k}^0 c (\cos\beta - \cos\alpha \cos\gamma)/\sin\gamma$$

$$- S_{32} x_{1k}^0 a \cos\gamma - S_{32} x_{2k}^0 b - S_{32} x_{3k}^0 c \cos\alpha + x_{3j} - x_{3j}^0 - S_{31} x_{1k}$$

$$+ S_{31} x_{1k}^0 - S_{32} x_{2k} + S_{32} x_{2k}^0 - S_{33} x_{3k} + S_{33} x_{3k}^0 \quad . \qquad (2.26)$$

The molecular rotations and translations affect the X_j. We wish to rotate the molecule about axis ℓ, where $|\ell| = 1$. The components of ℓ are the direction cosines of the rotation axis with respect to the reference Cartesian system defined by matrix $\underline{D}$. There will be a set of three orthogonal rotation axes; these may be taken parallel to the Cartesian axes, or, for example, parallel to the molecular inertial axes. The effect of a rotation of θ about ℓ is given by the matrix equation [2.28]

$$\underline{X}_j' = \underline{R}(\theta, \underline{\ell}) \left(\underline{X}_j - \underline{X}_j^0 \right) + \underline{X}_j^0 \qquad (2.27)$$

where

$$\underline{R} = \begin{bmatrix} \cos\theta + \ell_1^2(1 - \cos\theta) & \ell_1\ell_2(1 - \cos\theta) + \ell_3 \sin\theta & \ell_1\ell_3(1 - \cos\theta) - \ell_2 \sin\theta \\ \ell_1\ell_2(1 - \cos\theta) - \ell_3 \sin\theta & \cos\theta + \ell_2^2(1 - \cos\theta) & \ell_2\ell_3(1 - \cos\theta) + \ell_1 \sin\theta \\ \ell_1\ell_3(1 - \cos\theta) + \ell_2 \sin\theta & \ell_2\ell_3(1 - \cos\theta) - \ell_1 \sin\theta & \cos\theta + \ell_3^2(1 - \cos\theta) \end{bmatrix} . \qquad (2.28)$$

Inspection of (2.24-26) shows that there are no interaction terms between rotations and the lattice constants. This is also true for the translations, which are simply

$$\underline{X}_j' = \underline{X}_j + \Delta\underline{X}_j \quad , \qquad (2.29)$$

where again the subscript on $\Delta\underline{X}$ allows for more than one molecule in the asymmetric unit.

The expressions for the effect of a rotation θ about ℓ for the components of $\underline{C}$ are

$$C_i = R_{i1}(X_{1j}-X_{1j}^0) + R_{i2}(X_{2j}-X_{2j}^0) + R_{i3}(X_{3j}-X_{3j}^0) + X_{ij}^0$$

$$- S_{i1}\left[R_{11}(X_{1k}-X_{1k}^0)+R_{12}(X_{2k}-X_{2k}^0)+R_{13}(X_{3k}-X_{3k}^0)+X_{1k}^0\right]$$

$$- S_{i2}\left[R_{21}(X_{1k}-X_{1k}^0)+R_{22}(X_{2k}-X_{2k}^0)+R_{23}(X_{3k}-X_{3k}^0)+X_{2k}^0\right]$$

$$- S_{i3}\left[R_{31}(X_{1k}-X_{1k}^0)+R_{32}(X_{2k}-X_{2k}^0)+R_{33}(X_{3k}-X_{3k}^0)-X_{3k}^0\right] - T_i \quad . \tag{2.30}$$

Similar expressions for the effect of a translation along $\underline{\ell}$ are

$$C_i = X_{ij} + \ell_i \Delta X_{ij} - S_{i1}(X_{1k}+\ell_1\Delta X_{1k})$$

$$- S_{i2}(X_{2k}+\ell_2\Delta X_{2k}) - S_{i3}(X_{3k}+\ell_3\Delta X_{3k}) - T_i \quad . \tag{2.31}$$

2.3.3 The Structural Derivatives of $\underline{C}$

As indicated by (2.11), we will need the structural first derivatives of $\underline{C}$. If the Newton-Raphson method is used for structural refinement, we will also need the second derivatives. In addition, for the optimization of the nonbonded potential parameters we will need derivatives of the forces and torques with respect to the potential parameters. All of these derivatives can be evaluated numerically. However, since a lattice sum is involved, numerical differentiation can be expensive in computer time.

It is possible (although somewhat tedious) to evaluate all of these derivatives analytically. The advantages obtained with the use of analytical derivatives are greater accuracy and speed of evaluation. Section 2.7 gives the analytical derivatives of C_1, C_2, and C_3 with respect to the lattice constants. Section 2.7 is similar to Table 1 in [2.28] except that the results have been extended to apply to the case of nonorigin rotations.

To consider the derivatives of $\underline{C}$ with respect to molecular rotation, we note that rotation operations about different axes do not commute. The effect of two rotations (among others) is given by

$$\underline{X}_j = \underline{R}_\mu\underline{R}_\nu\left(\underline{X}_j-\underline{X}_0^\nu\right) + \underline{R}_\mu\left(\underline{X}_0^\nu-\underline{X}_0^\mu\right) + \underline{X}_0^\mu \quad , \tag{2.32}$$

where $\underline{R}_\nu$ and $\underline{X}_0^\nu$ are the first rotation operator and the first rotation center, respectively. Since $\underline{S}$ and $\underline{T}$ are not affected by the rotation, the derivatives of $\underline{C}$ are

$$\frac{\partial C_i}{\partial \theta} = \frac{\partial X_{ij}'}{\partial \theta} - S_{i1}\frac{\partial X_{1k}'}{\partial \theta} - S_{i2}\frac{\partial X_{2k}'}{\partial \theta} - S_{i3}\frac{\partial X_{3k}'}{\partial \theta} \quad . \tag{2.33}$$

For two successive rotations, the second derivatives are

$$\frac{\partial^2 C_i}{\partial\theta_\mu\partial\theta_\nu} = \frac{\partial^2 X_{ij}'}{\partial\theta_\mu\partial\theta_\nu} - S_{i1}\frac{\partial^2 X_{1k}'}{\partial\theta_\mu\partial\theta_\nu} - S_{i2}\frac{\partial^2 X_{2k}'}{\partial\theta_\mu\partial\theta_\nu} - S_{i3}\frac{\partial^2 X_{3k}'}{\partial\theta_\mu\partial\theta_\nu} \quad . \tag{2.34}$$

The general form of the derivatives is complicated, but much simplification occurs when they are evaluated at the starting point, $\theta = 0$. For the first derivative we have

$$\frac{\partial X'_j}{\partial \theta} = \left[\frac{\partial R}{\partial \theta}\right]_0 (X_j - X^0) + X^0 \quad .$$

From (2.27) the derivative matrix at $\theta = 0$ is

$$\left[\frac{\partial R}{\partial \theta}\right]_0 = \begin{bmatrix} 0 & \ell_3 & -\ell_2 \\ -\ell_3 & 0 & \ell_1 \\ \ell_2 & -\ell_1 & 0 \end{bmatrix} \quad . \tag{2.35}$$

The second derivatives for rotation about the same axis are

$$\frac{\partial X'}{\partial \theta^2} = \left[\frac{\partial^2 R}{\partial \theta^2}\right]_0 (X_j - X^0) + X^0 \quad , \tag{2.36}$$

where

$$\left[\frac{\partial^2 R}{\partial \theta^2}\right]_0 = \begin{bmatrix} -1+\ell_1^2 & \ell_1\ell_2 & \ell_1\ell_3 \\ \ell_1\ell_2 & -1+\ell_2^2 & \ell_2\ell_3 \\ \ell_1\ell_3 & \ell_2\ell_3 & -1+\ell_3^2 \end{bmatrix} \quad . \tag{2.37}$$

If the rotation axes are different, we have

$$\frac{\partial^2 X'_j}{\partial \theta_\mu \partial \theta_\nu} = \left[\frac{\partial R_\mu}{\partial \theta_\mu}\right]_0 \left[\frac{\partial R_\nu}{\partial \theta_\nu}\right]_0 (X_j - X^0) \quad , \tag{2.38}$$

where

$$\left[\frac{\partial R_\mu}{\partial \theta_\mu}\right]_0 \left[\frac{\partial R_\nu}{\partial \theta_\nu}\right]_0 = \begin{bmatrix} -\ell_{3\mu}\ell_{3\nu} -\ell_{2\mu}\ell_{2\nu} & \ell_{2\mu}\ell_{1\nu} & \ell_{3\mu}\ell_{1\nu} \\ \ell_{1\mu}\ell_{2\nu} & -\ell_{3\mu}\ell_{3\nu} -\ell_{1\mu}\ell_{1\nu} & \ell_{3\mu}\ell_{2\nu} \\ \ell_{1\mu}\ell_{3\nu} & \ell_{2\mu}\ell_{3\nu} & -\ell_{2\mu}\ell_{2\nu} -\ell_{1\mu}\ell_{1\nu} \end{bmatrix} \quad , \tag{2.39}$$

and the second rotation center, X^0_μ, does not appear. It is seen that (2.39) is equal to (2.37) when $\mu = \nu$.

The results for the rotational derivatives may be conveniently summarized in matrix form if we define the first derivative matrix (2.35) as

$$L_\nu = \left[\frac{\partial R_\nu}{\partial \theta_\nu}\right]_0 \quad . \tag{2.40}$$

Then we have

$$\frac{\partial C}{\partial \theta} = L(X_j - X_0) - SL(X_k - X_0) \quad , \tag{2.41}$$

and

$$\frac{\partial^2 C}{\partial \theta_\mu \partial \theta_\nu} = \underline{L}_\mu \underline{L}_\nu (\underline{X}_j - \underline{X}_0^\nu) - \underline{SL}_\mu \underline{L}_\nu (\underline{X}_k - \underline{X}_0^\nu) \quad . \tag{2.42}$$

In (2.42) the lower and upper off-diagonal triangular parts of the second derivative matrix refer to forward and reverse sequences, respectively, of the rotation
matrices $\underline{R}_\mu$. The final symmetric second derivative matrix is obtained by expansion
of the triangular part which corresponds to the arbitrary choice of either forward
or reverse sequences of $\underline{R}_\mu$.

Finally, we need the derivatives of $\underline{C}$ with respect to the translation $\Delta\underline{X}$. The
first derivatives are

$$\frac{\partial \underline{C}}{\partial \Delta X_i} = \underline{\ell} - \underline{S\ell} \quad , \tag{2.43}$$

and the second derivatives are zero.

2.3.4· The Forces and Torques as Functions of the Nonbonded Potential Parameters

The first derivatives shown in (2.11) need to be evaluated. From (2.10) we obtain

$$\frac{\partial V}{\partial \underline{p}} = \sum_{j=1}^{N} \sum_{k}^{\infty} \frac{\partial V_{jk}(r,q)}{\partial \underline{p}} \quad . \tag{2.44}$$

For the empirical potential of (2.6) we have

$$\frac{\partial V_{jk}}{\partial \underline{p}} = \left[6Ar_{jk}^{-7} - BC \exp(-Cr_{jk}) - q_j q_k r_{jk}^{-2} \right.$$
$$\left. + 2p\beta D \left\{ \exp[-2\beta(r_e - r_{jk})] - \exp[-\beta(r_e - r_{jk})] \right\} \right] \left(\frac{\partial r_{jk}}{\partial \underline{p}} \right) \quad . \tag{2.45}$$

The last derivative may be expressed in terms of the components C_i of r_{jk}:

$$|C| = \left(C_1^2 + C_2^2 + C_3^2 \right)^{\frac{1}{2}} \quad , \tag{2.46}$$

$$\frac{\partial r_{jk}}{\partial \underline{p}} = \frac{1}{r_{jk}} \left[C_1 \frac{\partial C_1}{\partial \underline{p}} + C_2 \frac{\partial C_2}{\partial \underline{p}} + C_3 \frac{\partial C_3}{\partial \underline{p}} \right] \quad . \tag{2.47}$$

If the Newton-Raphson method is used for energy minimization, the second structural derivatives are also needed:

$$\frac{\partial^2 V_{jk}}{\partial p_\mu \partial p_\nu} = \left[6Ar_{jk}^{-7} - BC \exp(-Cr_{jk}) - q_j q_k r_{jk}^{-2} \right.$$
$$+ 2p\beta D \left\{ \exp[-2\beta(r_e - r_{jk})] - \exp[-\beta(r_e - r_{jk})] \right\} \left] \left(\frac{\partial^2 r_{jk}}{\partial p_\mu \partial p_\nu} \right) \right.$$
$$+ \left[-42Ar_{jk}^{-8} + BC^2 \exp(-Cr_{jk}) + q_j q_k r_{jk}^{-3} \right.$$
$$\left. + 2p\beta^2 D \left\{ 2\exp[-2\beta(r_e - r_{jk})] - \exp[-\beta(r_e - r_{jk})] \right\} \right] \left(\frac{\partial r_{jk}}{\partial p_\mu} \right) \left(\frac{\partial r_{jk}}{\partial p_\nu} \right) \quad . \tag{2.48}$$

the second derivatives of r_{jk} are

$$
\frac{\partial^2 r_{jk}}{\partial p_\mu \partial p_\nu} = \frac{1}{r_{jk}} \left[C_1 \frac{\partial^2 C_1}{\partial p_\mu \partial p_\nu} + C_2 \frac{\partial^2 C_2}{\partial p_\mu \partial p_\nu} + C_3 \frac{\partial^2 C_3}{\partial p_\mu \partial p_\nu} \right.
$$
$$
\left. + \left(\frac{\partial C_1}{\partial p_\mu}\right)\left(\frac{\partial C_1}{\partial p_\nu}\right) + \left(\frac{\partial C_2}{\partial p_\mu}\right)\left(\frac{\partial C_2}{\partial p_\nu}\right) + \left(\frac{\partial C_3}{\partial p_\mu}\right)\left(\frac{\partial C_3}{\partial p_\nu}\right) - \left(\frac{\partial r_{jk}}{\partial p_\mu}\right)\left(\frac{\partial r_{jk}}{\partial p_\nu}\right) \right] \quad . \qquad (2.49)
$$

A second type of second derivative is needed if the goal is to find the minimum forces and torques as a function of the potential parameters. These derivatives show the rate of change of the forces and torques with respect to changes in the potential parameters. These derivatives may be obtained by straightforward differentiation of (2.44) and (2.45) with respect to the potential parameters; note that (2.46) and (2.47) are independent of the potential parameters.

The potential parameters A, B, and p are of linear form in (2.45), and q is bi-linear. We will see in the following section that it may be appropriate to convert A and B to bilinear form by defining $a^2 \equiv A$ and $b^2 \equiv B$ and treating a and b as the variables. The nonlinear form of the potential parameters, C, β, and r_e makes them more difficult to derive from observed structures, as compared to a, b, and q.

2.4 Accelerated Convergence of Lattice Sums

The repulsive and the partial-bonding interactions are short-range, such that a discrete sum may be made over all contributing interactions. The Madelung (Coulombic) interaction is long range, and the dispersion interaction is intermediate in range. The problems encountered in the accurate evaluation of the lattice energy of an ionic crystal have long been known. For certain simple type structures, Madelung constants are available; these constants enable the evaluation of the Madelung energy as a function of the lattice constants or the nearest-neighbor distances. If the ionic crystal structure is not of a simple type for which a Madelung constant is available, this procedure obviously fails.

It would be possible to tabulate Madelung-type constants for the dispersion energy also. This is normally not done because the structures for which the dispersion energy is needed are seldom of a simple lattice type.

2.4.1 Fourier Transforms of the Lattice Sum

EWALD [2.29] showed how the Madelung energy lattice sum could be rapidly evaluated, based on a mathematical method originated by EPSTEIN [2.30]. The method was given a more general treatment by BERTAUT [2.31], for the Madelung sums. NIJBOER and DEWETTE [2.32] showed how the same technique could be used for the dispersion sum for simple lattice types. This technique of rapid evaluation of the dispersion sum has been extended to crystals of any general structure [2.33].

The basic idea involved in the accelerated convergence treatment of lattice sums is the conversion of a portion of the lattice sum by a Fourier transformation to reciprocal space. The apportionment of the sum between direct and reciprocal space is selected so that each part converges more rapidly (i.e., with fewer terms) than the direct space sum alone.

The general formulation of the direct space sum is

$$S_n = \frac{1}{2} \sum_{j \neq k} q_j q_k r_{jk}^{-n} \quad , \tag{2.50}$$

where the sum j is over one unit cell, and the sum k is over the lattice, except for the self-terms $j = k$ in the zeroth (origin) unit cell. The q_j are generalized coefficients; in the case $n = 1$ they correspond to net point charges. A convergence function $\phi(r)$ is selected such that $\phi(0) = 1$ and the function decreases rapidly with increasing r. Each term of S_n is multiplied by ϕ:

$$S_n = \frac{1}{2} \sum_{j \neq k} q_j q_k r_{jk}^{-n} \phi(r_{jk}) + \frac{1}{2} \sum_{j \neq k} q_j q_k r_{jk}^{-n} [1 - \phi(r_{jk})] \quad . \tag{2.51}$$

The first set of terms in (2.51) shows improved convergence because of the properties of ϕ. The convergence property of the second set of terms is not improved. The second set of terms will be transformed to reciprocal space; the reciprocal space sum can then be made to converge rapidly.

Thus the three-dimensional Fourier transform of the second sum in (2.51) is needed. Details of how to do this are given in [2.32]. The function $\phi(r)$ is chosen to be the normalized incomplete gamma function:

$$\phi(r) = \frac{\Gamma(n/2, K^2 \pi r^2)}{\Gamma(n/2)} = \frac{1}{\Gamma(n/2)} \int_{K^2 \pi r^2}^{\infty} t^{n/2-1} \exp(-t) dt \quad , \tag{2.52}$$

where the constant K determines how much of the direct sum is converted to reciprocal space. If $K = 0$, $\phi(r) = 1$ and no conversion occurs; as K increases, more of the direct sum is transformed to reciprocal space.

The Fourier transformation yields a sum over the reciprocal space variable, $\underline{h}$. There result also two important constant terms which are independent of both r and $\underline{h}$. If we define[1] $a^2 = \pi K^2 r_{jk}^2$ and $b^2 = \pi h_\lambda^2 / K^2$, the modified lattice sum is

$$S_n = \frac{1}{2\Gamma(n/2)} \left\{ \sum_{j \neq k} q_j q_k r_{jk}^{-n} \Gamma(n/2, a^2) + \frac{n^{-3/2}}{V_{cell}} \sum_{\underline{h} \neq 0} |F(\underline{h}_\lambda)|^2 h_\lambda^{n-3} \Gamma(-n/2 + 3/2, b^2) \right.$$
$$\left. + \frac{2\pi^{n/2} K^{n-3}}{(n-3)V_{cell}} \left[\sum_{cell} q_j \right]^2 - \frac{2\pi^{n/2} K^n}{n} \left[\sum_{cell} q_j^2 \right] \right\} \quad . \tag{2.53}$$

---

1 The use of a and b here refers only to the convergence acceleration procedure. The nonbonded parameters a and b are readily identifiable in their context. V_{cell} is the volume of the unit cell.

The sum converges for $n > 3$ or for $n > 0$ if $\sum_{cell} q_j = 0$.

The two cases of immediate interest are for $n = 1$ (Madelung sum) and $n = 6$ (dispersion sum). The formulation of S_n implicitly assumed the geometric mean combining law for the A_{jk}, the coefficients of the dispersion terms. If we define $a^2 = A$, $b^2 = B$, and $2c = C$, we have

$$A_{jk} = a_j a_k = \sqrt{A_{jj} A_{kk}} \quad , \tag{2.54}$$

$$B_{jk} = b_j b_k = \sqrt{B_{jj} B_{kk}} \quad , \tag{2.55}$$

$$C_{jk} = c_j + c_k = (1/2)(C_{jj} + C_{kk}) \quad , \tag{2.56}$$

which are the geometric mean combining rules. While the geometric mean combining law holds exactly for Madelung interactions, the accuracy of the law needs to be verified for V_d and V_r. A very great reduction in the number of nonbonded potential parameters occurs if (2.54,56) are sufficiently accurate for use in practical calculations.

The final convergence accelerated expressions for V_c and V_d are given below. The incomplete gamma functions have been converted to exponentials and the readily available complementary error function; $erfc(x)$.

$$erfc(x) = \int_x^\infty \exp(-t^2)dt \quad , \tag{2.57}$$

$$(V_c)_{total} = \frac{1}{2} \sum_{j \neq k} q_j q_k r_{jk}^{-1} erfc(a) + \frac{1}{2\pi V_{cell}} \sum_{\underline{h}_\lambda \neq 0} |F(\underline{h}_\lambda)|^2 |\underline{h}_\lambda|^{-2}$$

$$\times \exp(-b^2) - K \sum_{cell} q_j^2 \quad , \tag{2.58}$$

$$(V_d)_{total} = \frac{1}{2} \sum_{j \neq k} q_j q_k r_{jk}^{-6}(1 + a^2 + a^4/2)\exp(-a^2)$$

$$+ \frac{\pi^{9/2}}{3V_{cell}} \sum_{\underline{h}_\lambda \neq 0} |F(\underline{h}_\lambda)|^2 h_\lambda^3 \left[\pi^{\frac{1}{2}} erfc(b) + [1/(2b^3) - 1/b]\exp(-b^2)\right]$$

$$+ \frac{\pi^3 K^3}{6V_{cell}} \left(\sum_{cell} q_j\right)^2 - \frac{\pi^3 K^6}{12} \left(\sum_{cell} q_j^2\right) \quad . \tag{2.59}$$

These equations refer to V_c or V_d for all nonidentical atom pairs. In molecular crystals it is usually appropriate to subtract the intramolecular part, since this part includes bonded interactions as well as intramolecular nonbonded interactions. As far as the direct space sum is concerned, the intramolecular terms can simply be left out. The corresponding terms in the reciprocal space sum are always present and must be evaluated explicitly and subtracted. The correction term is evaluated in direct space:

$$\sum_{intra} q_j q_k r_{jk}^{-6} [1 - \phi(r_{jk})] \quad . \tag{2.60}$$

In addition, if the direct sum includes pair interactions to only one molecule when there are Z molecules per cell, the constant involving $\left(\sum q_j\right)^2$ needs to be multiplied by Z and the reciprocal lattice sum needs to be divided by Z. The energy is then calculated per mole rather than per unit cell.

2.4.2 Numerical Examples

Tables 2.1,2 are presented to give the reader a feel for the actual applications of accelerated convergence. Table 2.1 shows results for the Madelung energy of sodium chloride. Note first that the untreated sum is wildly in error. In practice, this indicates that some kind of convergence treatment is always required for Madelung sums. In spite of this fact, papers still appear occasionally in the literature in which direct Madelung sums are used. Since the Ewald method has been available for over a half-century, this situation should be corrected [2.34].

The last column in the table gives the relative speed of evaluation. Note that this speed is maximum around K = 0.3. It may be desirable to neglect the reciprocal sum because of the difficulty of evaluation of the analytical derivatives of that sum. The table shows that with a lower value of K, the reciprocal sum can be neglected while still retaining a fairly good speed of evaluation.

Table 2.2 shows the application of accelerated convergence to the dispersion energy of benzene. In this more complex calculation the accuracy targets are lowered. Also, this calculation differs from the previous one in that moderate accuracy can be achieved without convergence acceleration. But note again that no reasonable truncation limit will give high accuracy. Therefore, accelerated convergence should always be used if high accuracy is desired.

Table 2.1. Accelerated convergence results for sodium chloride Madelung energy [kJ/mol,Å]

K	Truncation limit Direct	Reciprocal	Sum Direct	Reciprocal	Total energy	Relative speed
(a) High accuracy (~0.005%)						
0.0	20.0	0.0	-2480.17	0.00	-2480.17	--
0.1	20.0	0.0	-861.04	0.00	-861.04	0.06
0.2	12.0	0.4	-861.06	1.01	-861.04	0.21
0.3	8.0	0.6	-923.22	62.20	-861.04	0.37
0.4	6.0	0.8	-1125.12	264.10	-861.04	0.27
(b) Moderate accuracy (~0.05%)						
0.1	8.0	0.0	-861.06	0.00	-861.06	0.11
0.2	6.0	0.4	-861.02	1.01	-861.01	0.57
0.3	6.0	0.4	-923.21	62.19	-861.04	1.00
0.4	6.0	0.6	-1125.12	264.10	-861.04	0.54

Table 2.2. Accelerated convergence results for benzene dispersion energy [kJ/mol,Å]

K	Truncation limit Direct	Reciprocal	Direct	Reciprocal	Energy	Speed
(a) High accuracy (~0.05%)						
0.0	18.0	0.0	-40.97	0.0	-40.97	--
0.1	14.0	0.0	-41.37	0.00	-41.37	0.17
0.2	8.0	0.3	-41.05	-0.32	-41.37	0.49
0.3	6.0	0.5	-22.86	-18.52	-41.38	0.59
0.4	6.0	0.7	114.22	-155.61	-41.39	0.26
(b) Moderate accuracy (~1%)						
0.0	18.0	0.0	-40.97	0.0	-40.97	0.08
0.1	10.0	0.0	-41.01	0.0	-41.01	0.50
0.2	8.0	0.0	-41.05	0.0	-41.05	1.00
0.3	6.0	0.4	-22.86	-18.46	-41.32	0.90
0.4	6.0	0.6	114.22	-155.47	-41.25	0.38

At moderate accuracy, accelerated convergence speeds up the calculation by more than an order of magnitude. With $K = 0.2$, the reciprocal sum can be neglected completely. Thus the accelerated convergence technique may be incorporated into computer programs to result in large savings of computer run time.

2.4.3 Lattice Energy Derivatives Including Accelerated Convergence

Equation (2.10) may be expanded using the results of the accelerated convergence method given above; for the asymmetric unit, $(V_c)_{total}$ plus $(V_d)_{total}$ is

$$\frac{1}{2} \sum_{jk} \left[-Ar_{jk}^{-6}\left(1+a_6^2+a_6^4/2\right) \exp\left(-a_6^2\right) + q_j q_k r_{jk}^{-1} \mathrm{erfc}(a_1) \right]$$

$$+ \frac{\pi^{9/2}}{3VZ} \sum_{\underset{\neq 0}{h_\lambda}} |F_6(h_\lambda)|^2 h_\lambda^3 \left[\pi^{\frac{1}{2}} \mathrm{erfc}(b_6) + \left(1/(2b_6^3)-1/b_6\right) \exp\left(-b_6^2\right) \right]$$

$$+ \frac{1}{2\pi VZ} \sum_{\underset{\neq 0}{h_\lambda}} |F_1(h_\lambda)|^2 h_\lambda^{-2} \exp\left(-b_1^2\right)$$

$$- \frac{\pi^3 K_6^3}{6V_{cell}Z} \left(\sum_{asym} A_{jj}^{\frac{1}{2}} \right)^2 + \frac{\pi^3 K_6^6}{12} \left(\sum_{asym} A_{jj} \right) - K_1 \left(\sum_{asym} q_j^2 \right)$$

$$+ \sum_{intra} \left\{ Ar_{jk}^{-6}\left(-a_6^2-a_6^4/2\right) \exp\left(-a_6^2\right) - q_j q_k r_{jk}^{-1}[1-\mathrm{erfc}(a)] \right\} \quad . \tag{2.61}$$

If the convergence constants K_1 and K_6 are small, the reciprocal lattice sums are small and do not vary much with the direct space structural parameters. Therefore we can restrict the analytical derivatives to the direct lattice sums, the convergence constants, and the intramolecular correction. If the molecule is rigid, the latter correction also does not vary; and only one of the convergence constants depends on structure through the cell volume. The first structural derivatives are

$$\frac{1}{2} \sum_{jk} \left\{ \left[Ar_{jk}^{-7} \left(6 + 6a_6^2 + 3a_6^4 + 2a_6^6 \right) \exp\left(-a_6^2 \right) - q_j q_k r_{jk}^{-2} \mathrm{erfc}(a_1) \right] \frac{\partial r_{jk}}{\partial p} \right.$$

$$\left. + q_j q_k r_{jk}^{-1} \frac{\partial \mathrm{erfc}(a_1)}{\partial p} \right\} + \frac{\pi^3 K_6^3}{6 Z V_{cell}^2} \left(\sum_{asym} A_{jj}^{\frac{1}{2}} \right) \frac{\partial V_{cell}}{\partial p} \quad . \tag{2.62}$$

The values of $\partial r/\partial p$ were considered above in Sect.2.3.3. The structural derivatives of the unit cell volume are also needed. Of course, $\partial V/\partial p = 0$ if p is a molecular rotation or translation. The first and second derivatives of V may be evaluated from (2.15) and are given in [2.28].

The second structural derivatives including accelerated convergence are analogous to (2.48) and are also given in [2.28]. The mixed potential-structural second derivatives are obtained from (2.62) by straightforward differentiation.

2.5 Derivation of Nonbonded Potential Parameters

The objective is to find the optimum numerical coefficients for the assumed form of the potential energy of the crystal. Some of the potential parameters are better defined by crystal structure data than others. For instance, there is a large correlation between B and C. If a somewhat less than optimum C is chosen, B can be adjusted to give an overall fit to the structure that is nearly as good as before. Also, the linear parameters such as B are the easiest to optimize. For these reasons we will assume reasonable values for C and not vary them (Sect.2.2.2).

The lattice energy equations take an especially simple form if only A, B, q, and p are optimized and if the geometric mean combining law is assumed. For a given type of atom-pair interactions (α,β) the lattice sum can be factored out as independent of the potential parameters

$$V^{\alpha\beta} = -a_j a_k \sum_{\alpha\beta} r_{jk}^{-6} + b_j b_k \sum_{\alpha\beta} \exp(-Cr_{jk}) + q_j q_k \sum_{\alpha\beta} r_{jk}^{-1} + p_j p_k \sum_{\alpha\beta} V_m \quad . \tag{2.63}$$

2.5.1 The Force-Fit Method

At the observed equilibrium structure the lattice energy is at a minimum and all forces and torques are zero. We define a function to be minimized

$$G = \sum w \left(\frac{\partial V}{\partial p} \right)^2 + \sum w' (V - V_0)^2 \quad , \tag{2.64}$$

where the first sum is over the forces and torques and the second sum is over the lattice energies of the various structures. V_0 is the estimated static lattice energy obtained by correcting the observed heat of sublimation to 0 K and for zero point energy.

It is necessary to include at least one term in the lattice energy to normalize
the depth of the potential wells. In the absence of such a constraint, the variation
of the potential parameters could yield $V = 0$ everywhere to satisfy the condition of
minimum forces and torques. The second sum may be viewed as a side condition introduced as a penalty function into G. The weight w' is adjusted to be just large enough
to give the desired fit to V_0. Often many of the V_0 are not known, or are subject to
large errors. The most accurate values of V_0 are obtained by direct calorimetry.
Measurements of vapor pressure using the Clausius-Clapeyron treatment yield much
less accurate values for V_0. In practice, one or two good values for V_0 may be chosen
as sufficient for normalization purposes. It should be remarked that no merit what-
ever is shown for potential parameters merely on the basis that they reproduce an
energy to which they are normalized.

The function G may be minimized by linearized least squares, as a function of the
potential parameters. The energy, forces, and torques are expanded in a Taylor's
series about trial values for the parameters. This procedure is iterated until the
minimum is found.

The refined values for the potential parameters are sensitive to the choice of
weights w. Unit weighting is not appropriate or satisfactory. A simple expansion of
the force about the trial model point is

$$F_p - F_p^0 = \Delta p \, \frac{\partial F}{\partial p} = \Delta p \, \frac{\partial^2 V}{\partial p^2} \quad . \tag{2.65}$$

If we associate $F_p - F_p^0$ with $\sigma(F)$ and Δp with $\sigma(p)$, we obtain the following relation
between the standard deviations:

$$\sigma(F) = \sigma(p) \, \frac{\partial^2 V}{\partial p^2} \quad . \tag{2.66}$$

The value for the weighting factor is

$$w = [\sigma(F)]^{-2} \quad . \tag{2.67}$$

To use (2.66) we need to estimate $\sigma(p)$. Some reasonable estimates are: unit cell
edge 1%, cell angles and molecular rotations 0.02 rad, and molecular translations
0.05 Å [2.35].

The above described scheme for obtaining weighting factors makes use of only the
diagonal elements of the Hessian matrix of the crystal energy. For hydrocarbons, at
least, this diagonal weighting scheme appears to be satisfactory.

BUSING [2.36] describes a full matrix weighting scheme which makes use of the
complete Hessian rather than just the diagonal elements. BATES and BUSING [2.37] have
used this full matrix weighting method to successfully derive nonbonded potential
parameters from the crystal structure of hexachlorobenzene. HSU and WILLIAMS [2.38]
showed that the diagonal weighting method does not work well for the derivation of
nonbonded potential parameters from the crystal structure of hexachlorobenzene.

Thus it appears that although the diagonal weighting scheme may be satisfactory in many cases, the full matrix weighting scheme may be more universal in its applicability. The necessity for use of the full matrix weighting scheme is probably related to the presence of strong correlations in the crystal data and in the potential energy model parameters.

2.5.2 Application to Hydrocarbons

The substances in a hydrocarbon data base are shown in Fig.2.1. There are 18 crystal structures of which half are aromatic, yielding 118 forces and torques to be fitted.

Calculation with this data base lead to the following conclusions:

a) the geometric mean combining law holds well for the dispersion energy [2.14,15],
b) the Madelung (Coulombic) energy in hydrocarbons is significant [2.16],

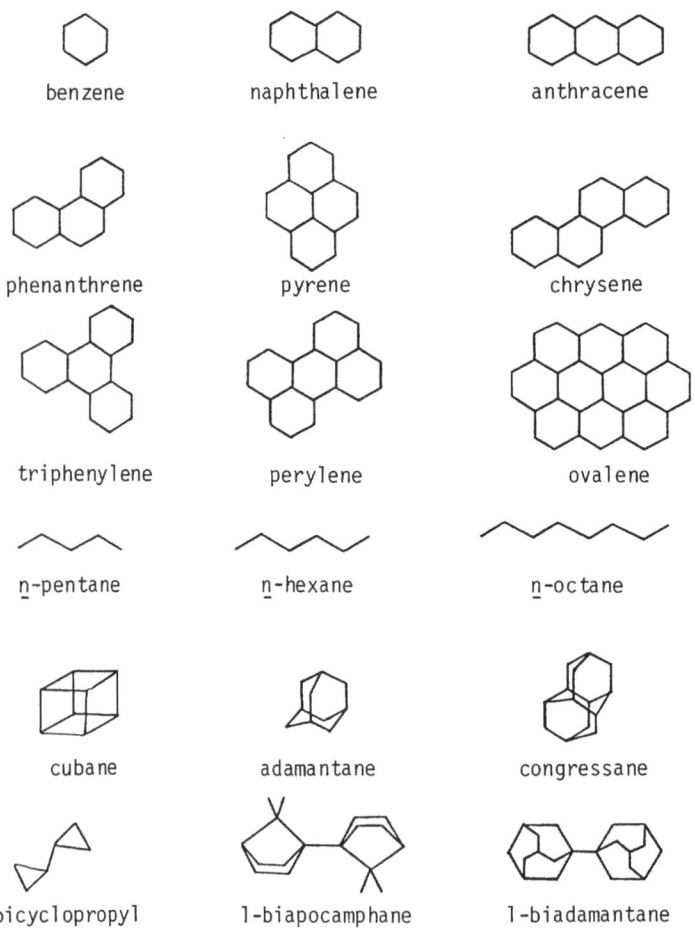

benzene   naphthalene   anthracene

phenanthrene   pyrene   chrysene

triphenylene   perylene   ovalene

n-pentane   n-hexane   n-octane

cubane   adamantane   congressane

bicyclopropyl   1-biapocamphane   1-biadamantane

Fig. 2.1. A hydrocarbon crystal structure data base suitable for the derivation of (exp-6-1) nonbonded potential parameters [2.39]

c) the geometric mean combining law holds well for the repulsion energy provided the Madelung energy is included [2.16],

d) there does not appear to be a significant difference between aromatic and saturated carbon V_d and V_r [2.11],

e) the structure-derived potentials yield good prediction of lattice vibrational frequencies [2.35],

f) lattice energies (heats of sublimation) can be predicted rather well even though not specifically fitted [2.16],

g) the structural parameters of the fitted structures are predicted with about the error assumed in the weights listed above,

h) the force-fit method gives more transferable parameters than the direct-parameter-fit method (see below),

i) the (exp-6-1) model yields a better fit with less calculation effort than the EPEN potential [2.39].

Very recently there has appeared a detailed analysis of the atom-atom (exp-6-1) empirical nonbonded potential for ethylene by WASIUTYNSKI, VAN DER AVOIRD, and BERNS [2.4]. These authors calculated the intermolecular potential and lattice dynamics of ethylene by ab initio quantum mechanics. Their analysis showed that the empirical model is in good agreement with the ab initio calculations. They confirm the importance of electrostatic interactions, which are represented in the empirical model by Coulomb's law operating between point net charges.

The ab initio results do show some specific deviations (e.g., anisotropy) from the empirical model. No doubt, such quantum-mechanical calculations will become possible for larger molecules in the future. The calculations may indicate the desirability of introducing changes in the empirical functional forms. These improved empirical nonbonded potentials can be used for larger molecules such as proteins, etc. It is unlikely that ab initio quantum-mechanical calculations for molecules as large as proteins can be made in the near future. Thus, empirical potentials are expected to continue to be useful for large molecules in the near future. Figure 2.2 illustrates the form of the (exp-6) part of the empirical nonbonded potentials for hydrogen and carbon nonbonded interactions. These potentials were derived from the crystal structures of the molecules shown in Fig.2.1. These basis structures were chosen with the intention that the nonbonded potentials be transferable to hydrocarbon molecules in general and to hydrocarbon portions of molecules. Of course, no such empirical nonbonded potentials can be completely accurate because of the approximations inherent in the model. The experience with hydrocarbons has shown that the following nonbonded parameters for carbon and hydrogen are reliable for many purposes [2.39] (units are kJ/mol, Å, and e)

A_{HH} = 136 B_{HH} = 11677 C_{HH} = 3.74   (C-H bond foreshortening = 0.07)

A_{CC} = 2414 B_{CC} = 367250 C_{CC} = 3.60   (q_C in benzene C-H group = 0.153) .

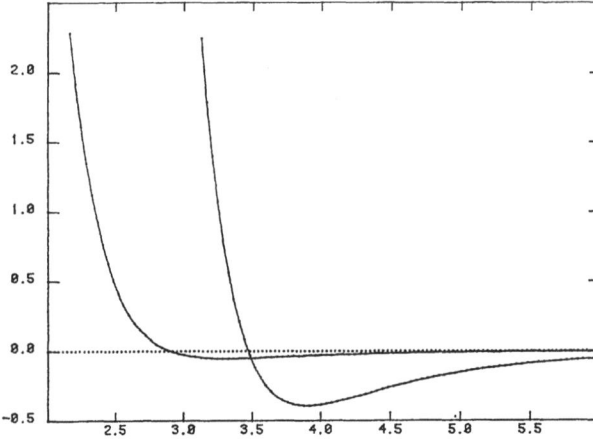

2.5.3 The Direct-Parameter-Fit Method

The idea of this method is to predict the structural parameter shifts for a given
set of potential parameters [2.40]. Expansion of the forces (or torques) about a
trial point yields

$$F_i = F_i^0 + \sum_j \Delta p_j \frac{\partial F_i}{\partial p_j} + \frac{1}{2} \sum_j \sum_k \Delta p_j \Delta p_k \frac{\partial^2 F_i}{\partial p_j \partial p_k} + \ldots \quad . \tag{2.66}$$

At equilibrium $\underline{F} = 0$, and by neglecting the higher-order terms we obtain

$$0 = \underline{F}^0 + \underline{\underline{H}} \Delta \underline{p} \quad , \tag{2.67}$$

or

$$\Delta \underline{p} = -\underline{\underline{H}}^{-1} \underline{F}^0 \quad . \tag{2.68}$$

We can then replace the first sum of (2.64):

$$G' = [\sigma^2(p)]^{-2} (\Delta p)^2 + w'(V - V_0)^2 \quad . \tag{2.69}$$

In order to minimize G' it is necessary to evaluate $\partial \Delta p / \partial q$. This cannot be done
analytically because of the dependence on the inverse of the Hessian matrix $\underline{\underline{H}}$. Numerical differentiation can be used however.

  The results obtained with (2.69) are plausibly close to the results obtained with
(2.64). The fits obtained to the observed structures, however, are not as good as
with the force-fit method [2.39]. In addition, the force-fit method yields better
predictions of the lattice vibrational frequencies [2.35]. The only obvious reason
for the failure of the direct-parameter-fit method is that the higher-order terms
in (2.66) must be significant for this application.

2.6 Extensions of the Potential Model

The simplifying assumptions of molecular rigidity and zero thermal motion have been made. Many interesting large molecules (e.g., proteins) can be treated as nearly rigid, except for certain allowed rotations about intramolecular bonds [2.6]. The potential energy model can be modified to allow free or hindered rotation about such bonds [2.41].

The lattice vibrational frequencies can be calculated fairly well from the structure-derived potential parameters [2.35]. The calculation of thermal expansion requires some inclusion of the effects of thermal motion on the lattice constants.

2.6.1 Molecular Rotations and Translations

A molecule, especially one that has a nearly round shape, can rotate in the crystal by overcoming the rotational energy barrier presented by nonbonded interactions. The simplest model is to hold the observed crystal structure constant while rotating a single reference molecule about the appropriate axis. Such calculations have been performed [2.42,43] and reasonable agreement was obtained with experimental data. A more elaborate model would allow some motion of the neighboring molecules while the reference molecule is rotating.

Translational and rotational vibrations are discussed in Sect.2.6.4 below. The extreme case of molecular translation is molecular self-diffusion. Models for self-diffusion using empirical nonbonded potential functions have been formulated and tested (for example, [2.44]). Disordered crystal structures may be analyzed in terms of nonbonded interactions (for example, [2.45]). The study of substitutional polymorphism is an interesting field in itself. Here a different molecule, which has somewhat similar nonbonded interactions, is substituted at a molecular site in the crystal [2.1].

2.6.2 Intramolecular Rotations

During a rotation about an intramolecular bond the intramolecular energy will vary. For a single bond joining two saturated carbon atoms, a barrier of threefold symmetry is expected. For rotation about a single bond joining two ethylenic or aromatic carbon atoms, a twofold rotational barrier is expected. In addition to these valence electronic effects, the intramolecular nonbonded energy will vary with the change in molecular conformation. It is usual to define a torsion potential [2.1] as a function of the bond rotation ϕ:

$$V(\phi) = (V_0/2)(1 + \cos n\phi) \quad . \tag{2.68}$$

The torsion angle ϕ is measured from the eclipsed conformation, so that $V(\phi)$ is defined to be zero at the staggered conformation. The value of V_0 for ethane, for instance, is known to be about 12 kJ/mol.

Molecules having a biphenyl-type bond will show a twofold barrier because of the possibility of conjugation in the planar conformation. On the other hand, the nonbonded interactions between the groups *ortho* to the phenyl-phenyl bond are maximum in the planar conformation. Biphenyl itself is not a good case for study since the molecule is planar, or nearly so, in the crystal. When substituents are present on the rings, the phenyl rings may become nonplanar in the crystal. It is then possible to study the balance between nonbonded intramolecular interactions, intermolecular interactions, and the torsion potential.

An interesting example is the crystal structure of p,p'-bitolyl [2.46]. In this compound the *para* methyl groups destroy the uniform thickness of the phenyl groups. Therefore, the efficient molecular packing of planar unsubstituted biphenyl molecules cannot be obtained. In fact, there are two molecules of p,p'-bitolyl in the asymmetric unit which have different torsion angles. Since the molecules are chemically identical, the two different torsion angles must result from the effects of intermolecular nonbonded energy.

The torsion potential was derived from the p,p'-bitolyl structure as follows [2.28]. The conjugation potential was assumed as in (2.68) with $n = 2$ and the optimum value of V_0 was found by fitting the twelve degrees of rotational and translational freedom of the molecules. V_0 was found to be 9 kcal/mol, a figure which is in good agreement with theoretical estimates made by quantum-mechanical calculations.

It is especially interesting that the molecular packing analysis correctly predicts the two different twist angles for the two molecules in the asymmetric unit. This is a clear instance of success in calculating the effects of intermolecular forces on molecular conformation in considerable detail.

Another interesting facet of this calculation is that the predicted gas phase torsion angle is smaller than the angles observed in the crystal. Thus, p,p'-bitolyl may be contrasted with biphenyl in going from gas to crystal. In the former molecule the torsion angle increases, while in the latter molecule the torsion angle decreases to zero in the crystal.

2.6.3 Thermal Corrections

The correct equilibrium structure at constant pressure and temperature of a crystal is one with minimum free energy, rather than simply minimum potential energy (ΔV^* represents a change in volume here):

$$\Delta G = \Delta V + P \Delta V^* + T \Delta S \quad . \tag{2.70}$$

For crystals near atmospheric pressure the $P\Delta V^*$ term is negligible; the $T\Delta S$ term, while small, may not always be neglected. For instance, there are numerous examples of the existence of different crystal structures of the same substance at different temperatures; these are thermal polymorphs.

The primary structural effect of thermal motion is a general slight expansion, which may be quite anisotropic and therefore structure dependent. A secondary effect is that there are small changes in the structure itself in terms of the rotational and translational positions of the molecules in the unit cell. The most detailed calculations will consider the populations of the lattice-vibrational modes (see the next section). We describe here a much simpler method of treating thermal expansion and thermal molecular orientation changes [2.22].

The nonbonded interatomic potential may be expanded in a Taylor series in the displacement variable $u = r - r_e$:

$$V = V(r_e) + uV'(r_e) + \frac{1}{2} u^2 V''(r_e) + \frac{1}{6} u^3 V'''(r_e) + \ldots \quad . \tag{2.71}$$

At equilibrium the mean value of the force is zero, which leads to the result [2.22]:

$$\frac{dV}{dr} = 0 = <u>V''(r_e) + \frac{1}{2} <u^2>V'''(r_e) \tag{2.72}$$

or

$$<u> = K<u^2> \quad , \quad \text{where} \quad K = -V'''(r_e)/2V''(r_e) \quad .$$

The mean displacement is thus related to the mean square displacement through the anharmonicity factor K.

The (exp-6) potentials may be altered to reflect this displacement by increasing B sufficiently to shift the minimum of the potential by $<u>$. The new value of B is

$$B' = B[r_e/(r_e+<u>)]^7 \exp(C<u>) \quad . \tag{2.73}$$

A separate alteration can be made for each B as is appropriate for anisotropic $<u^2>$. For rigid molecules a good approximation to $<u^2>$ is given by [2.47]

$$<u^2> = \underline{\ell}^t \underline{T} \underline{\ell} + (\underline{\ell} \times \underline{r})^t \underline{L} (\underline{\ell} \times \underline{r}) \quad , \tag{2.74}$$

where $\underline{\ell}$ is a unit vector in the direction of the displacement, $\underline{r}$ is an atomic position relative to the center of libration, and $\underline{T}$ and $\underline{L}$ are the translational and librational thermal tensors. Experimental values of $\underline{T}$ and $\underline{L}$ are available from X-ray diffraction studies, or they may be estimated theoretically. The mean displacement is then readily calculated from the mean-square displacement with (2.72).

Table 2.3. Thermal correction and thermal expansion for crystalline argon [K,Å]

T	$<u^2>$	Lattice constant	
		Observed	Calculated
0	0.0117	5.311	5.303
20	0.0150	5.318	5.314
40	0.0243	5.348	5.348
60	0.0361	5.393	5.393

Table 2.4. Thermal correction and thermal expansion for anthracene

T	Observed				Calculated			
	a	b	c	β	a	b	c	β
95	8.443	6.002	11.124	125.6	8.164	6.023	11.016	124.3
290	8.562	6.038	11.184	124.7	8.354	6.101	11.135	123.9
change	0.119	0.036	0.060	-0.9	0.190	0.078	0.119	-0.4

Table 2.3 shows how well this procedure works for crystalline argon. At the two higher temperatures the fit to the observed thermal expansion is excellent. At the lower temperatures the lattice constants are calculated too small, indicating that other factors need to be considered when the total thermal motion is small. Table 2.4 shows the observed and calculated thermal expansion of anthracene, using this method. The anisotropy of expansion is correctly reproduced, with the smallest expansion occurring along b and the largest along a. The direction of change in β, the cell angle, is also correct. The change in molecular orientation at the two temperatures was also calculated [2.22] and found to be in fair agreement with the observed reorientation.

2.6.4 Lattice Vibrations

The introduction of thermal vibrations into the lattice necessitates further consideration of several factors. The first and most obvious is the question of possible structural changes caused by the amplitudes of vibration, such as the thermal expansion discussed in the previous section. Further, it is observed experimentally that the lattice frequencies vary with temperature. In the quasi-harmonic approximation [2.48] the temperature variation is the indirect result of thermal expansion.

If one is to allow translational and rotational vibrations of the molecule, the rigidity of the molecule needs to be considered. For benzene the errors in the calculated frequencies due to the rigid-molecule assumption are small [2.49]. For slightly less rigid molecules such as naphthalene [2.50] or durene [2.51] errors reach about 10% because of the neglect of intramolecular nonrigidity. Nevertheless, the rigid-molecule approximation is useful if one does not expect accuracy better than about 10% or so.

In making a calculation of lattice frequencies it is important that first the lattice energy be minimized to insure that all forces and torques are zero. Failure

to do this preliminary step will result in errors in the calculated lattice frequencies [2.35]. Also, this procedure will ensure that the nonbonded potentials
predict the static structure fairly well.

It is possible in some cases to derive nonbonded parameters from lattice-frequency observations alone. A much better procedure is to combine lattice-frequency
data with structural data by the force-fit method. For a set of 18 hydrocarbons with
118 structural observational equations, when 58 observed lattice frequencies were
added, essentially no change was obtained in the nonbonded potentials [2.35]. Thus,
the structure-derived nonbonded potentials were already nearly optimized in this
case for the calculation of lattice frequencies. On the other hand, if the direct-
parameter-fit method was used, the fit to the frequencies was not as good.

The overall fit to the 58 observed lattice frequencies of the five hydrocarbon
crystals benzene, naphthalene, anthracene, phenanthrene, and pyrene is reported in
[2.35]. A r.m.s. error of fit of 26.8 cm^{-1} was obtained for the direct-parameter-fit
Coulombic model, 19.6 cm^{-1} for the force noncoulombic model, and 13.6 cm^{-1} for the
force Coulombic model. The force Coulombic model yields the best overall fit. In
fact, the fit with this model is as good as can be expected without some way of
handling the temperature variation of the lattice frequencies. The structural data
alone do not show such a marked preference of the force Coulombic model over the
direct-parameter-fit model [2.39].

2.6.5 Weak Morse Bond Potential

There exists a class of (nominally) molecular crystals in which larger-than-usual
intermolecular forces are present, and in which hydrogen bonding is absent. The
nomenclature of these crystals is not well developed at present. Many of the cases
of fairly strong interactions are referred to as charge-transfer complexes [2.52,53].
Many crystal structure determinations [2.54] of these molecular complexes indicate
that they involve specific atom-atom intermolecular distances which are shorter than
those normally expected from the sum of the van der Waals radii of the atoms. Such
a weak bond could have a potential energy similar to a bond in a diatomic molecule
as represented by the Morse potential (2.5).

Consider the example of crystalline chlorine. Its structure is different from
the other diatomic solids N_2, O_2, and F_2 in a way that cannot be understood with an
(exp-6) nonbonded potential [2.55]. The Cl_2 molecules pack in approximately hexagonal
layers in the ab plane with the molecules tilted $32°$ from the perpendicular to the
plane [2.56]. The structure of these sheets can be understood from normal nonbonded
interactions. However, the intersheet packing cannot be so understood. The normal
expectation is that the end of a chlorine molecule in one sheet would fit into the
center of a triangle of atoms on the next sheet. Instead of doing this, the chlorine
is placed at the center of a line joining two atoms in the adjacent sheet. Attempts
to explain this structure with normal nonbonded potentials fail.

Table 2.5. Deviation of the calculated structure models of minimum potential energy from the observed structure of chlorine [Å,deg]

Model	Δa	Δb	Δc	$\Delta\theta$
V_d+V_r	-1.09	-1.26	0.50	11.9
$V_d+V_r+V_q$	-0.10	-0.67	1.44	30.7
$V_d+V_r+V_m$	0.01	-0.63	1.12	16.7
$V_d+V_r+V_q+V_m$	0.01	-0.01	0.01	0.1

The abnormal intersheet geometry described above is accompanied by very short Cl-Cl distances of 3.32 Å. The next shortest nonbonded distances are 3.74 Å. Thus, these very short distances seem to involve a specific interaction that might be represented by a Morse potential. YAMASAKI [2.57] suggested this procedure in 1962, but did not carry a full calculation with the Morse potential. Table 2.5 shows how well the chlorine crystal structure can be fitted with and without a Morse potential [2.58,59]. V_q represents a molecular electronic quadrupole by placing net atomic charges on the chlorines and the molecular center.

The crystal structure is orthorhombic, space group Cmca [2.56], so that the crystal structure variables are the lattice constants $\underline{a}$, $\underline{b}$, and $\underline{c}$. The molecules lie on a crystallographic mirror plane at a twofold (2/m site symmetry) and can be described by the tilt angle θ. The columns in Table 2.5 show the differences between the observed and calculated structures.

It is seen that the addition of either the quadrupole or the Morse potential improves the agreement. Much better agreement is obtained if both the quadrupole and the Morse potential are allowed.

The Raman stretching frequency of Cl_2 is 557 cm^{-1} in the gas phase, which shifts to 538 cm^{-1} in the crystal at 77 K [2.60]. This shift is consistent with the presence of a Morse interaction. The shifts for Br_2 and I_2 are larger and one can predict that stronger Morse bonds are present in those crystal structures. Also, the gaseous chlorine dimer $(Cl_2)_2$ has a detectable dipole moment [2.61], a fact which is consistent with a significant Morse interaction.

2.6.6 High-Pressure Crystal Structures

The compressibility of crystals is generally small. A covalently bonded crystal such as diamond will have a smaller change in its lattice constants with increasing applied pressure than that of a molecular crystal such as benzene. In addition to changes in the lattice constants, a structural transition may occur. Ice, for example, has many known high-pressure polymorphs.

Crystalline benzene also provides an example of a pressure-induced phase transition. The normal crystal structure at atmospheric pressure and low temperature is orthorhombic, space group Pbca, with four molecules in the unit cell [2.62]. At

25 kbar pressure the crystal structure is monoclinic, with two molecules in the unit cell [2.63]. It should be possible to understand the relative stabilities of the orthorhombic and monoclinic crystal structures through the use of empirical non-bonded potential functions for the clamped crystal.

At zero pressure the forces and torques in the crystal are zero as discussed above (atmospheric pressure is negligible in its effect in this situation). A uniform hydrostatic pressure can be modelled for the crystal by setting the three derivatives of the lattice energy with respect to the cell edges (forces) equal in magnitude to the applied pressure. In other words, instead of targeting these forces to zero, they are targeted to the value of the applied pressure. The molecular rotation angles in the cell must also be allowed to change. No translation of the molecule is involved since the molecular center resides in a crystallographic inversion center in both structure types.

The results of this type of treatment do show the correct order of stabilities [2.64] for the benzene polymorphs, provided net atomic charges are included in the nonbonded potential model. The necessity for the use of the slightly more complex (exp-6-1) potential here, rather than a noncoulombic (exp-6) potential illustrates a point. The point is that although a nonbonded potential of modest accuracy may suffice in some instances, a more accurate potential is needed to distinguish more subtle phenomena such as this example.

A further refinement of the treatment of the high-pressure phase transition in crystalline benzene includes the $P\Delta V^*$ terms. The addition of this term does not change the predicted order of stabilities, which is still correct [2.65].

2.7 Appendix. Derivatives of $\underline{C}$

Listed below are the first and second derivatives of $\underline{C}$ with respect to the lattice constants. The second derivatives that are not listed are equal to zero. This list is an extension of the expressions given in [2.28].

$$\frac{\partial C_1}{\partial a} = -\left(t_1 - x_{1j}^0 + S_{11}x_{1k}^0\right)\sin\gamma - S_{12}x_{1k}^0 \cos\gamma$$

$$\frac{\partial C_1}{\partial b} = -S_{12}x_{2k}^0$$

$$\frac{\partial C_1}{\partial c} = -\left(t_3 - x_{3j}^0 + S_{11}x_{3k}^0\right)(\cos\beta - \cos\alpha \cos\gamma)/\sin\gamma - S_{12}x_{3k}^0 \cos\alpha - S_{13}x_{3k}^0 \delta/\sin\gamma$$

$$\frac{\partial C_1}{\partial \alpha} = -\left(t_3 - x_{3j}^0 + S_{11}x_{3k}^0\right)(c \sin\alpha \cos\gamma/\sin\gamma)$$
$$+ S_{12}x_{3k}^0 c \sin\alpha - S_{13}x_{3k}^0 c \sin\alpha(\cos\alpha - \cos\beta \cos\gamma)/(\delta \sin\gamma)$$

$$\frac{\partial C_1}{\partial \gamma} = -\left(t_1 - x_{1j}^0 + S_{11}x_{1k}^0\right)a \cos\gamma + \left(t_3 - x_{3j}^0 + S_{11}x_{3k}^0\right)c(\cos\beta \cos\gamma - \cos\alpha)/\sin^2\gamma$$
$$+ S_{13}x_{3k}^0 \ c[\delta \cos\gamma/\sin^2\gamma + (\cos\alpha \cos\beta - \cos\gamma)/\delta] + S_{12}x_{1k}^0 \ a \sin\gamma$$

$$\frac{\partial C_2}{\partial a} = -\left(t_1 - x_{1j}^0 + S_{22}x_{1k}^0\right)\cos\gamma - S_{21}x_{1k}^0 \ \sin\gamma$$

$$\frac{\partial C_2}{\partial b} = -\left(t_2 - x_{2j}^0 + S_{22}x_{2k}^0\right)$$

$$\frac{\partial C_2}{\partial c} = -\left(t_3 - x_{3j}^0 + S_{22}x_{3k}^0\right)\cos\alpha - S_{21}x_{3k}^0(\cos\beta - \cos\alpha \cos\gamma)/\sin\gamma - S_{23}x_{3k}^0\delta/\sin\gamma$$

$$\frac{\partial C_2}{\partial \alpha} = \left(t_3 - x_{3j}^0 + S_{22}x_{3k}^0\right)c \sin\alpha - S_{21}x_{3k}^0 \ c \sin\alpha \cos\gamma/\sin\gamma$$
$$- S_{23}x_{3k}^0 \ c \sin\alpha(\cos\alpha - \cos\beta \cos\gamma)/(\delta \sin\gamma)$$

$$\frac{\partial C_2}{\partial \beta} = S_{21}x_{3k}^0 \ c \sin\beta/\sin\gamma - S_{23}x_{3k}^0 \ c \sin\beta(\cos\beta - \cos\alpha \cos\gamma)/(\delta \sin\gamma)$$

$$\frac{\partial C_2}{\partial \gamma} = \left(t_1 - x_{1j}^0 + S_{22}x_{1k}^0\right)a \sin\gamma - S_{21}x_{1k}^0 \ a \cos\gamma$$
$$+ S_{21}x_{3k}^0 \ c(\cos\beta \cos\gamma - \cos\alpha)/\sin^2\gamma + S_{23}x_{3k}^0 \ c[\delta \cos\gamma/\sin^2\gamma + (\cos\alpha \cos\beta - \cos\gamma)/\delta]$$

$$\frac{\partial C_3}{\partial a} = - S_{31}x_{1k}^0 \ \sin\gamma - S_{32}x_{1k}^0 \ \cos\gamma$$

$$\frac{\partial C_3}{\partial b} = - S_{32}x_{2k}^0$$

$$\frac{\partial C_3}{\partial c} = -\left(t_3 - x_{3j}^0 + S_{33}x_{3k}^0\right)(\delta/\sin\gamma) - S_{31}x_{3k}^0(\cos\beta - \cos\alpha \cos\gamma)/\sin\gamma - S_{32}x_{3k}^0 \ \cos\alpha$$

$$\frac{\partial C_3}{\partial \alpha} = \left(t_3 - x_{3j}^0 + S_{33}x_{3k}^0\right)c \sin\alpha(\cos\beta \cos\gamma - \cos\alpha)/(\delta \sin\gamma)$$
$$- S_{31}x_{3k}^0 \ c \sin\alpha \cos\gamma/\sin\gamma + S_{32}x_{3k}^0 \ c \sin\alpha$$

$$\frac{\partial C_3}{\partial \beta} = \left(t_3 - x_{3j}^0 + S_{33}x_{3k}^0\right)c \sin\beta(\cos\alpha \cos\gamma - \cos\beta)/(\delta \sin\gamma) + S_{31}x_{3k}^0 \ c \sin\beta/\sin\gamma$$

$$\frac{\partial C_3}{\partial \gamma} = \left(t_3 - x_{3j}^0 + S_{33}x_{3k}^0\right)c[(\cos\alpha \cos\beta - \cos\gamma)/\delta + \delta \cos\gamma/\sin^2\gamma] - S_{31}x_{1k}^0 a \cos\gamma$$
$$+ S_{31}x_{3k}^0 \ c(\cos\beta \cos\gamma - \cos\alpha)/\sin^2\gamma - S_{32}x_{1k}^0 a \sin\gamma$$

$$\frac{\partial^2 C_1}{\partial \gamma \partial a} = -\left(t_1 - x_{1j}^0 + S_{11}x_{1k}^0\right)\cos\gamma + S_{12}x_{1k}^0 \ \sin\gamma$$

$$\frac{\partial^2 C_1}{\partial a \partial c} = -\left(t_3 - x_{3j}^0 + S_{11}x_{3k}^0\right)\sin a \cos\gamma/\sin\gamma$$
$$+ S_{12}x_{3k}^0 \ \sin\alpha - S_{13}x_{3k}^0 \ \sin\alpha(\cos\alpha - \cos\beta \cos\gamma)/(\delta \sin\gamma)$$

$$\frac{\partial^2 C_1}{\partial \beta \partial c} = \left(t_3 - x_{3j}^0 + S_{11}x_{3k}^0\right)\sin\beta/\sin\gamma - S_{13}x_{3k}^0 \sin\beta(\cos\beta - \cos\alpha \cos\gamma)/(\delta \sin\gamma)$$

$$\frac{\partial^2 C_1}{\partial \gamma \partial c} = \left(t_3 - x_{3j}^0 + S_{11}x_{3k}^0\right)(\cos\beta \cos\gamma - \cos\alpha)/\sin^2\gamma$$
$$+ S_{13}x_{3k}^0[\delta \cos\gamma/\sin^2\gamma + (\cos\alpha \cos\beta - \cos\gamma)/\delta]$$

$$\frac{\partial^2 C_1}{\partial \alpha^2} = -\left(t_3 - x_{3j}^0 + S_{11}x_{3k}^0\right)c \cos\alpha \cos\gamma/\sin\gamma + S_{12}x_{3k}^0 c \cos\alpha$$
$$+ \left(S_{13}x_{3k}^0 c/\sin\gamma\right)[\sin^2\alpha + \cos\alpha \cos\beta \cos\gamma - \cos^2\alpha)/\delta + \sin^2\alpha(\cos\alpha - \cos\beta \cos\gamma)^2/\delta^3]$$

$$\frac{\partial^2 C_1}{\partial \beta \partial \alpha} = \left(S_{13}x_{3k}^0 c \sin\alpha \sin\beta/\sin\gamma\right)[(\cos\alpha - \cos\beta \cos\gamma)(\cos\beta - \cos\alpha \cos\gamma)/\delta^3 - \cos\gamma/\delta]$$

$$\frac{\partial^2 C_1}{\partial \gamma \partial \alpha} = \left(t_3 - x_{3j}^0 + S_{11}x_{3k}^0\right)c \sin\alpha/\sin^2\gamma - S_{13}x_{3k}^0 c \sin\alpha \cos\beta/\delta$$
$$+ S_{13}x_{3k}^0 c \sin\alpha(\cos\alpha - \cos\beta \cos\alpha)[\cos\gamma/(\delta \sin^2\gamma) + (\cos\gamma - \cos\alpha \cos\beta)/\delta^3]$$

$$\frac{\partial^2 C_1}{\partial \beta^2} = \left(t_3 - x_{3j}^0 + S_{11}x_{3k}^0\right)c \cos\beta/\sin\gamma + S_{13}x_{3k}^0 c \sin^2\beta/(\delta \sin\gamma)$$
$$+ \left(S_{13}x_{3k}^0 c/\sin\gamma\right)(\cos\beta - \cos\alpha \cos\gamma)[\sin^2\beta(\cos\beta - \cos\alpha \cos\gamma)/\delta^3 - \cos\beta/\delta]$$

$$\frac{\partial^2 C_1}{\partial \gamma \partial \beta} = -\left(t_3 - x_{3j}^0 + S_{11}x_{3k}^0\right)c \sin\beta \cos\gamma/\sin^2\gamma - S_{13}x_{3k}^0 c \cos\alpha \sin\beta/\delta$$
$$+ S_{13}x_{3k}^0 c \sin\beta(\cos\beta - \cos\alpha \cos\gamma)[\cos\gamma/(\delta \sin^2\gamma) + (\cos\gamma - \cos\alpha \cos\beta)/\delta^3]$$

$$\frac{\partial^2 C_1}{\partial \gamma^2} = \left(t_1 - x_{1j}^0 + S_{11}x_{1k}^0\right)a \sin\gamma - \left(t_3 - x_{3j}^0 + S_{11}x_{3k}^0\right)c[\cos\beta(1 + \cos^2\gamma) - 2\cos\alpha \cos\gamma]/\sin^3\gamma$$
$$+ S_{13}x_{3k}^0 c[1 - \cos\alpha \cos\beta \cos\gamma)/(\delta \sin\gamma) - \delta(\cos^2\gamma + 1)/\sin^3\gamma$$
$$+ \sin\gamma(\cos\gamma - \cos\alpha \cos\beta)^2/\delta^3] + S_{12}x_{1k}^0 a \cos\gamma$$

$$\frac{\partial^2 C_2}{\partial \gamma \partial a} = \left(t_1 - x_{1j}^0 + S_{22}x_{1k}^0\right)\sin\gamma - S_{21}x_{1k}^0 \cos\gamma$$

$$\frac{\partial^2 C_2}{\partial \alpha \partial c} = \left(t_3 - x_{3j}^0 + S_{22}x_{3k}^0\right)\sin\alpha - S_{21}x_{3k}^0 \sin\alpha \cos\gamma/\sin\gamma$$
$$- S_{23}x_{3k}^0 \sin\alpha(\cos\alpha - \cos\beta \cos\gamma)/(\delta \sin\gamma)$$

$$\frac{\partial^2 C_2}{\partial \beta \partial c} = S_{21}x_{3k}^0 \sin\beta/\sin\gamma - S_{23}x_{3k}^0 \sin\beta(\cos\beta - \cos\alpha \cos\gamma)/(\delta \sin\gamma)$$

$$\frac{\partial^2 C_2}{\partial \gamma \partial c} = S_{21}x_{3k}^0(\cos\beta \cos\gamma - \cos\alpha)/\sin^2\gamma + S_{23}x_{3k}^0[\delta \cos\gamma/\sin^2\gamma + (\cos\alpha \cos\beta - \cos\gamma)/\delta]$$

$$\frac{\partial^2 C_2}{\partial \alpha^2} = \left(t_3 - x_{3j}^0 + S_{22}x_{3k}^0\right)c \cos\alpha - S_{21}x_{3k}^0 c \cos\alpha \cos\gamma/\sin\gamma + \left(S_{23}x_{3k}^0 c/\sin\gamma\right)$$
$$\times [(\sin^2\alpha + \cos\alpha \cos\beta \cos\gamma - \cos^2\alpha)/\delta + \sin^2\alpha(\cos\alpha - \cos\beta \cos\gamma)^2/\delta^3]$$

$$\frac{\partial^2 C_2}{\partial\beta\partial\alpha} = \left(S_{23}x_{3k}^0\ c\ \sin\alpha\ \sin\beta/\sin\gamma\right)[(\cos\alpha-\cos\beta\ \cos\gamma)(\cos\beta-\cos\alpha\ \cos\gamma)/\delta^3-\cos\gamma/\delta]$$

$$\frac{\partial^2 C_2}{\partial\gamma\partial\alpha} = S_{21}x_{3k}^0\ c\ \sin\alpha/\sin^2\gamma\ -\ S_{23}x_{3k}^0\ c\ \sin\alpha\ \cos\beta/\delta$$
$$+\ S_{23}x_{3k}^0\ c\ \sin\alpha(\cos\alpha-\cos\beta\ \cos\gamma)[\cos\gamma/(\delta\ \sin^2\gamma)+(\cos\gamma-\cos\alpha\ \cos\beta)/\delta^3]$$

$$\frac{\partial^2 C_2}{\partial\beta^2} = S_{21}x_{3k}^0\ c\ \cos\beta/\sin\gamma\ +\ S_{23}x_{3k}^0\ c\ \sin^2\beta/(\delta\ \sin\gamma)$$
$$+\ \left(S_{23}x_{3k}^0\ c/\sin\gamma\right)(\cos\beta-\cos\alpha\ \cos\gamma)[\sin^2\beta(\cos\beta-\cos\alpha\ \cos\gamma)/\delta^3-\cos\beta/\delta]$$

$$\frac{\partial^2 C_2}{\partial\gamma\partial\beta} = -S_{21}x_{3k}^0\ c\ \sin\beta\ \cos\gamma/\sin^2\gamma\ -\ S_{23}x_{3k}^0\ c\ \cos\alpha\ \sin\beta/\delta$$
$$+\ S_{23}x_{3k}^0\ c\ \sin\beta(\cos\beta-\cos\alpha\ \cos\gamma)[\cos\gamma/(\delta\ \sin^2\gamma)+(\cos\gamma-\cos\alpha\ \cos\beta)/\delta^3]$$

$$\frac{\partial^2 C_2}{\partial\gamma^2} = \left(t_1-x_{1j}^0+S_{22}x_{1k}^0\right)a\ \cos\gamma\ +\ S_{21}x_{1k}^0\ a\ \sin\gamma$$
$$-\ S_{21}x_{3k}^0\ c[\cos\beta(1+\cos^2\gamma)-2\cos\alpha\ \cos\gamma]/\sin^3\gamma$$
$$+\ S_{23}x_{3k}^0\ c[(1-\cos\alpha\ \cos\beta\ \cos\gamma)/(\delta\ \sin\gamma)$$
$$-\delta(\cos^2\gamma+1)/\sin^3\gamma+\sin\gamma(\cos\gamma-\cos\alpha\ \cos\beta)^2/\delta^3]$$

$$\frac{\partial^2 C_3}{\partial\gamma\partial\alpha} = -S_{31}x_{1k}^0\ \cos\gamma\ +\ S_{32}x_{1k}^0\ \sin\gamma$$

$$\frac{\partial^2 C_3}{\partial\alpha\partial c} = \left(t_3-x_{3j}^0+S_{33}x_{3k}^0\right)\sin\alpha(\cos\beta\ \cos\gamma-\cos\alpha)/(\delta\ \sin\gamma)$$
$$-\ S_{31}x_{3k}^0\ \sin\alpha\ \cos\gamma/\sin\gamma\ +\ S_{32}x_{3k}^0\ \sin\alpha$$

$$\frac{\partial^2 C_3}{\partial\beta\partial c} = \left(t_3-x_{3j}^0+S_{33}x_{3k}^0\right)\sin\beta(\cos\alpha\ \cos\gamma-\cos\beta)/(\delta\ \sin\gamma)\ +\ S_{31}x_{3k}^0\ \sin\beta/\sin\gamma$$

$$\frac{\partial^2 C_3}{\partial\gamma\partial c} = \left(t_3-x_{3j}^0+S_{33}x_{3k}^0\right)[(\cos\alpha\ \cos\beta-\cos\gamma)/\delta+\delta\ \cos\gamma/\sin^2\gamma]$$
$$+\ S_{31}x_{3k}^0(\cos\beta\ \cos\gamma-\cos\alpha)/\sin^2\gamma$$

$$\frac{\partial^2 C_3}{\partial\alpha^2} = \left(t_3-x_{3j}^0+S_{33}x_{3k}^0\right)(c/\sin\gamma)[(\sin^2\alpha+\cos\alpha\ \cos\beta\ \cos\gamma-\cos^2\alpha)/\delta$$
$$+\sin^2\alpha(\cos\alpha-\cos\beta\ \cos\gamma)^2/\delta^3]\ -\ S_{31}x_{3k}^0\ c\ \cos\alpha\ \cos\gamma/\sin\gamma\ +\ S_{32}x_{3k}^0\ c\ \cos\alpha$$

$$\frac{\partial^2 C_3}{\partial\beta\partial\alpha} = \left(t_3-x_{3j}^0+S_{33}x_{3k}^0\right)(c\ \sin\alpha\ \sin\beta/\sin\gamma)[(\cos\alpha-\cos\beta\ \cos\gamma)$$
$$\times(\cos\beta-\cos\alpha\ \cos\gamma)/\delta^3-\cos\gamma/\delta]$$

$$\frac{\partial^2 C_3}{\partial\gamma\partial\alpha} = \left(t_3-x_{3j}^0+S_{33}x_{3k}^0\right)c\ \sin\alpha[-\cos\beta/\delta+\cos\gamma(\cos\alpha-\cos\beta\ \cos\gamma)/(\delta\ \sin^2\gamma)$$
$$+(\cos\alpha-\cos\beta\ \cos\gamma)(\cos\gamma-\cos\alpha\ \cos\beta)/\delta^3]\ +\ S_{31}x_{3k}^0\ c\ \sin\alpha/\sin^2\gamma$$

$$\frac{\partial^2 C_3}{\partial \beta^2} = \left(t_3 - x_{3j}^0 + S_{33}x_{3k}^0\right)(c/\sin\gamma)[\sin^2\beta/\delta + \sin^2\beta(\cos\beta - \cos\alpha \, \cos\gamma)^2/\delta^3$$
$$-\cos\beta(\cos\beta - \cos\alpha \, \cos\gamma)/\delta] + S_{31}x_{3k}^0 \, c \, \cos\beta/\sin\gamma$$

$$\frac{\partial^2 C_3}{\partial\gamma\partial\beta} = \left(t_3 - x_{3j}^0 + S_{33}x_{3k}^0\right)c[\sin\beta(\cos\beta - \cos\alpha \, \cos\gamma)(\cos\gamma - \cos\alpha \, \cos\beta)/\delta^3 - \cos\alpha \, \sin\beta/\delta$$
$$+\sin\beta \, \cos\gamma(\cos\beta - \cos\alpha \, \cos\gamma)/(\delta \, \sin^2\gamma)] - S_{31}x_{3k}^0 \, c \, \sin\beta \, \cos\gamma/\sin^2\gamma$$

$$\frac{\partial^2 C_3}{\partial\gamma^2} = \left(t_3 - x_{3j}^0 + S_{33}x_{3k}^0\right)c[(1 - \cos\alpha \, \cos\beta \, \cos\gamma)/(\delta \, \sin\gamma)$$
$$-\delta(\cos^2\gamma + 1)/\sin^3\gamma + \sin\gamma(\cos\gamma - \cos\alpha \, \cos\beta)^2/\delta^3] + S_{31}x_{1k}^0 \, a \, \sin\gamma$$
$$- S_{31}x_{3k}^0 \, c[\cos\beta(1 + \cos^2\gamma) - 2\cos\alpha \, \cos\gamma]/\sin^3\gamma + S_{32}x_{1k}^0 \, a \, \cos\gamma$$

Acknowledgement. This work was supported by National Institute of Health Research Grant GM16260.

References

2.1 A.I. Kitaigorodsky: *Molecular Crystals and Molecules* (Academic, New York, London 1973)
2.2 O. Kennard, D.G. Watson, F.H. Allen, S.M. Weeds: *Molecular Structures and Dimensions*, Vols.1-8 (International Union of Crystallography, Cambridge, England, 1970-1977)
2.3 N.L. Allinger: Adv. Phys. Org. Chem. *13*, 1 (1976); O. Ermer: Struct. Bonding (Berlin) *27*, 161 (1976)
2.4 T. Wasiutynski, A. van der Avoird, R.M. Berns: J. Chem. Phys. *69*, 5288 (1978)
2.5 R.E. Dickerson, I. Geis: *The Structure and Action of Proteins* (Harper & Row, New York, Evanston, London 1969)
2.6 D.A. Brant: Annu. Rev. Biophys. Bioeng. *1*, 369 (1972)
2.7 H. Margenau, N.R. Kestner: *Theory of Intermolecular Forces* (Pergamon, Oxford 1969)
2.8 J.O. Hirschfelder: *Intermolecular Forces*, in Advances in Chemical Physics, Vol.12 (Wiley-Interscience, New York 1967)
2.9 J.M. Parson, P.E. Siska, Y.T. Lee: J. Chem. Phys. *56*, 1511 (1972)
2.10 P.B. Foreman, P.K. Rol, K.P. Coffin: J. Chem. Phys. *61*, 1658 (1974)
2.11 D.E. Williams: Trans. Am. Crystallogr. Assoc. *6*, 21 (1970)
2.12 F. London: Z. Phys. Chem. (Leipzig) *B11*, 222 (1930); Trans. Faraday Soc. *33*, 8 (1937)
2.13 J.O. Hirschfelder, C.F. Curtiss, R.B. Bird: *Molecular Theory of Gases and Liquids* (John Wiley and Sons, New York 1964)
2.14 D.E. Williams: J. Chem. Phys. *45*, 3770 (1966)
2.15 D.E. Williams: J. Chem. Phys. *47*, 4680 (1967)
2.16 D.E. Williams: Acta Crystallogr. *A30*, 71 (1974)
2.17 B. Pullman (ed.): *Intermolecular Interactions: From Diatomics to Biopolymers* (Wiley, New York 1978); P. Hobza, R. Zahradnik: *Weak Intermolecular Interactions in Chemistry and Biology* (Elsevier, Amsterdam 1980)
2.18 A.D. Crowell: J. Chem. Phys. *29*, 446 (1958)
2.19 T.S. Kuan, A. Warshel, O. Schnepp: J. Chem. Phys. *52*, 3012 (1970)
2.20 Y.C. Lin: Ph.D. Dissertation, University of Louisville, Louisville, Kentucky (1973)

2.21 R.P. Rinaldi, G.S. Pawley: J. Phys. *C8*, 599 (1975)
2.22 D.E. Williams: Acta Crystallogr. *A28*, 84 (1972)
2.23 J.E. Jordan, E.A. Mason, I. Amdur: *Physical Methods of Chemistry*, Vol.I, Part IIID (Wiley-Interscience, New York 1972) p.365
2.24 S. Kita, K. Noda, H. Inouye: J. Chem. Phys. *64*, 3446 (1976)
2.25 R.F. Stewart, E.R. Davidson, W.T. Simpson: J. Chem. Phys. *42*, 3175 (1965)
2.26 O. Tapia, G. Bessis, S. Bratoz: Int. J. Quantum Chem. *S4*, 289 (1971)
2.27 T.L. Starr, D.E. Williams: J. Chem. Phys. *66*, 2054 (1977). The very recent configuration-interaction (CI) quantum-mechanical calculation of S.L. Price, A.J. Stone: Mol. Phys. *40*, 805 (1980) gave 0.17 Å for the repulsion center shift
2.28 D.E. Williams: Acta Crystallogr. *A28*, 629 (1972)
2.29 P.P. Ewald: Ann. Phys. (Leipzig) *64*, 253 (1921)
2.30 P. Epstein: Math. Ann. *56*, 615 (1903)
2.31 F. Bertaut: J. Phys. Radium *13*, 499 (1952)
2.32 B.R.A. Nijboer, F.W. Dewette: Physica *23*, 309 (1957)
2.33 D.E. Williams: Acta Crystallogr. *A27*, 452 (1971)
2.34 J.L. Derissen, P.H. Smit, J. Voogd: J. Phys. Chem. *81*, 1474 (1977)
2.35 T.L. Starr, D.E. Williams: Acta Crystallogr. *A33*, 771 (1977)
2.36 W.R. Busing: Trans. Am. Crystallogr. Assoc. *6*, 57 (1970)
2.37 J.B. Bates, W.R. Busing: J. Chem. Phys. *60*, 2414 (1974)
2.38 L.Y. Hsu, D.E. Williams: Acta Crystallogr. *A36*, 277 (1980)
2.39 D.E. Williams, T.L. Starr: Comput. Chem. *1*, 173 (1977)
2.40 A.T. Hagler, S. Lifson: Acta Crystallogr. *B30*, 1336 (1974)
2.41 F.A. Momany, R.F. McGuire, A.W. Burgess, H.A. Scheraga: J. Phys. Chem. *79*, 2361 (1975)
2.42 R.K. Boyd, C.A. Fyfe, D.A. Wright: J. Phys. Chem. Solids *35*, 1355 (1974)
2.43 C.A. Fyfe, D. Harold-Smith: Can. J. Chem. *54*, 769 (1976)
2.44 C.A. Fyfe, D. Harold-Smith: Can. J. Chem. *54*, 783 (1976)
2.45 J. Bernstein, K. Mirsky: Acta Crystallogr. *A34*, 161 (1978)
2.46 G. Casalone, C. Mariani, A. Magnoli, M. Simonetta: Mol. Phys. *15*, 339 (1968)
2.47 D.W.J. Cruickshank: Acta Crystallogr. *9*, 754 (1956)
2.48 G. Liebfried, W. Ludwig: Solid State Phys. *12*, 275 (1961)
2.49 G. Taddei, H. Bonadeo, M.P. Marzocchi, S. Califano: J. Chem. Phys. *58*, 966 (1973)
2.50 G.S. Pawley, S.J. Cyvin: J. Chem. Phys. *52*, 4073 (1970)
2.51 M. Sanquer, J.C. Messager: Mol. Cryst. Liq. Cryst. *29*, 285 (1975)
2.52 R.S. Mulliken, W.B. Person: *Molecular Complexes* (Wiley-Interscience, New York 1969)
2.53 R. Foster (ed.): *Molecular Association*, Vol.1 (Academic, New York 1975)
2.54 R. Foster: *Organic Charge-Transfer Complexes*, (Academic, New York 1969)
2.55 C.A. English, J.A. Venables: Proc. Roy. Soc. London *A340*, 57 (1974)
2.56 J. Donohue, S.H. Goodman: Acta Crystallogr. *18*, 568 (1965). A high precision refinement of the chlorine structure has been recently reported by E.D. Stevens: Mol. Phys. *37*, 27 (1979)
2.57 K. Yamasaki: J. Phys. Soc. Jpn. *17*, 1262 (1962)
2.58 L.Y. Hsu, D.E. Williams: Inorg. Chem. *18*, 79 (1979); D.E. Williams: Inorg. Chem. *19*, 2200 (1980)
2.59 L.Y. Hsu, D.E. Williams: unpublished research
2.60 A. Anderson: Chem. Phys. Lett. *6*, 611 (1970)
2.61 A.J. Harris, S.E. Novick, J.S. Winn, W. Klemperer: J. Chem. Phys. *61*, 3866 (1974)
2.62 G.E. Bacon, N. Curry, S.A. Wilson: Proc. Roy. Soc. London *A279*, 98
2.63 G.J. Piermarini, A.D. Mighell, C.E. Weir, S. Block: Science *165*, 1250
2.64 D. Hall, D.E. Williams: Acta Crystallogr. *A31*, 56 (1975)
2.65 D. Hall, T.H. Starr, D.E. Williams, M.K. Wood: Acta Crystallogr. *A36*, 494 (1980)
2.66 E. Scrocco, J. Tomasi: Adv. Quantum Chem. *11*, 115 (1978)
2.67 F.A. Momany: J. Phys. Chem. *82*, 592 (1978)
2.68 S.R. Cox, D.E. Williams: J. Comput. Chem., in press
2.69 H. Umeyama, K. Morokuma: J. Am. Chem. Soc. *99*, 1316 (1976)
2.70 P.H. Smit, J.L. Derissen, F.B. van Duijneveldt: Molec. Phys. *37*, 251 (1979)
2.71 S.R. Cox, L.Y. Hsu, D.E. Williams: Acta Crystallogr., in press

3. Conformational Analysis and Polypeptide Drug Design

F. A. Momany

With 21 Figures

The conformational energy program for peptides can compute with great success the conformational energy minima of peptides and other biologically interesting molecules. It uses empirically parametrized electrostatic and modified Lennard-Jones 12-6 potentials and, if needed, hydrogen bonding 12-10 potentials.

This program can aid experimental programs in drug design, and as is shown in detail for seven case studies, it can predict the biologically active conformers, thereby offering great savings in drug design research programs.

3.1 Introductory Comments

During the past ten years, computational studies on polypeptides have evolved from relatively simple hard-sphere approximations to studies using empirical and quantum-mechanical energy calculations. The emphasis has been to determine the low-energy conformations of small- to medium-sized polypeptides as well as to elucidate the type and magnitudes of interactions which lead to low-energy structures. Recent interest has focused on conformational energy calculations on polypeptide hormones and opiate analogs and the relationships between conformation and biological activity [3.1-11].

The lowest-energy conformation of a polypeptide is commonly thought of as being related to the minimum free energy structure, thus, it is reasonable to assume that a conformation of low energy should correspond closely to the conformation of highest population in solution. It is more difficult to understand the relationship between a calculated low-energy conformation of a molecule in the free state (vacuum) and the structure of, for example, a polypeptide hormone as it may exist while interacting with a membrane-bound receptor. To rationalize this relationship, one might assume that *one* of the low-energy "most probable" structures will satisfy the receptor configuration. If it is remembered that the lowest-energy free-state conformation may not necessarily be the structure best suited to fit a receptor, and that one must look for the conformation(s) that best fit the available analog-activity data, then it is possible to use the calculated conformations as guides for the design of new and more potent polypeptide hormones.

The strategies for polypeptide drug design using conformational information will be considered in this chapter. Major emphasis will be given only to small- or medium-sized open-chain polypeptides since the small cyclic molecules have structural constraints which limit their conformational adaptability. We will define medium-size polypeptides as those having five to fifteen amino acid residues.

This chapter will emphasize the conformational effects of structural modifications of polypeptides, and in particular the use of empirical conformational energy calculations as a guide to the design of hormone agonists and antagonists. Guidelines will emerge which may help initiate a new era in 'rational drug design' for polypeptides. Hopefully, using these guidelines, the design, synthesis, and testing of polypeptide analogs will be far more cost-effective than the traditional residue-by-residue substitution methods.

If these guidelines are to become practical and useful to medicinal chemists, protein chemists, pharmacologists, and others, then both the capabilities and the limitations imposed on the conformational aspects of polypeptide studies must be documented, and properly related to other pertinent physical, chemical and biological information.

One of the goals of this chapter is to provide the reader with a working knowledge of conformational aspects of polypeptides, and in particular the applications of conformational energy calculations. A complete review of all available literature on polypeptide conformations would perhaps be desirable. However, because so many overlapping and duplicating calculations exist, the author has severely constrained the literature citations to a few pertinent examples of each problem discussed.

In particular, most of the good and practical computational methods do yield the same or similar low-energy regions of conformational space for helical, bent, and extended structures. Therefore, except in unique situations, a complete bibliography becomes unnecessary. An apology is given here to those authors whose work is not explicitly referenced, together with the hope that no truly notable work has been overlooked or interpretation slighted.

3.2 Computational Procedures

Before specific conformational problems associated with polypeptide structure are discussed, it is necessary to outline the 'partitioned' potential energy procedure used by the author in computing the conformational energy of polypeptides [3.1]. In recent years it has become possible to use very sophisticated quantum-mechanical computations for conformational studies on structures the size of dipeptides and a few tripeptides [3.12,13]. However, it is still outside the range of computational ability to apply these methods to larger polypeptides. Further, since the results of molecular-orbital methods and the more recent 'empirical' methods [3.1] give nearly equivalent results as to positions of energy minima [3.9,14,15], the empirical calculations seem justified for use in the larger molecules to be discussed here.

'Partitioned' potential energy procedures have been reviewed elsewhere [3.16,17] and will be described only briefly here. The following ingredients are essential for the computational procedures: a) the structure (i.e., bond lengths and angles) of each amino acid residue, b) a computational method for generating any desired conformation of any sequence of amino acids, c) the empirical potential-energy parameters and functions for atom-atom interactions between all parts of the polypeptide chain for any conformation, d) an energy minimization routine which depends only on conformation, and finally, e) a method to include solvation and librational entropy contributions to the energy. It will not be possible here to describe current efforts to improve the potential functions, parameters [3.18-20], atomic charges [3.21], and computational techniques [3.5,6]. However, the current status of empirical calculations is such that one can confidently predict that calculated structures of low energy will closely represent those conformations allowed in solution. A major difficulty, which is yet to be solved, is to find the global energy minimum out of a manifold of many local minima.

The computer program used for empirical energy calculations by this author is designated ECEPP, for "Empirical Conformational Energy Program for Peptides." This program has been deposited with the Quantum Chemistry Program Exchange (QCPE) at Indiana University, together with documentation to aid users. The geometric data used in the program (atomic coordinates) were obtained from X-ray and neutron diffraction studies (bond lengths and angles) of amino acids and peptides. The program generates atomic coordinates for any desired conformation of a polypeptide by specifying the dihedral angles for rotation about bonds. The structural parameters are more fully described in Sect.3.3. After generation of a desired conformation, the program computes the energy of interaction between atoms from the Cartesian coordinates. The energy calculation includes all the interactions between atom pairs whose interatomic distance is allowed to vary upon a change in a dihedral angle. Torsional energy contributions are included for selected bonds, the energy parameters and periodic functions being determined from experimental torsional barriers of model compounds [3.22]. The nonbonded parameters were determined from calculations on crystal structures of model compounds [3.20].

The potential energy functions used to compute the interaction energy for an atom pair ij are the following.

1) Electrostatic (Coulombic):

$$U_{el} = 332.0 \ q_i q_j / Dr_{ij} \ , \tag{3.1}$$

where q_i and q_j are partial atomic charges in electron units, D is the dielectric constant, r_{ij} is the internuclear distance in Angstrom units, and 332.0 is the factor converting energy units from (electronic units)^2/Angstrom to kcal/mol.

2) Nonbonded (modified Lennard-Jones 12-6 potential):

$$U_{nb} = FA^{k\ell}/r_{ij}^{12} - C^{k\ell}/r_{ij}^6 \quad , \tag{3.2}$$

where the coefficients $A^{k\ell}$ and $C^{k\ell}$ are assigned specific values for each combination of atom types k and ℓ, and F has a different value, depending upon whether the type of interaction is $1{\to}4$ or $1{\to}5$ or greater [3.20,22].

3) Hydrogen bonded (substituted for (3.2) in appropriate cases):

$$U_{hb} = A_{Hx}/r_{Hx}^{12} - B_{Hx}/r_{Hx}^{10} \quad , \tag{3.3}$$

where A_{Hx} and B_{Hx} are specific coefficients for the different combinations of proton donors and oxygen or nitrogen acceptors [3.20].

In the calculations and discussions reported here, the nomenclature and conventions adopted by an IUPAC-IUB Commission were used [3.23]. Solvent effects are not included explicitly. Energy minimization is carried using procedures described elsewhere [3.1,5,6]. Reference to other computational methods will be given when results from other authors are discussed.

3.3 Polypeptides and Their Constituents

Theoretical as well as experimental studies on the conformational properties of polypeptides and their constituents have long been dominated by considering the amino acid as having "backbone" and "side-chain" components. The basic fragment of a polypeptide chain is denoted a "residue", as shown between solid brackets in Fig. 3.1. Another term which will be used regularly is "peptide unit". This unit involves the system of skeletal bonds from one α-carbon to the next and thus includes the amide fragment (-CONH-).

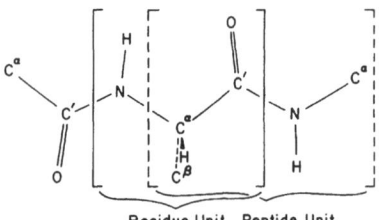

Residue Unit   Peptide Unit

Fig. 3.1. A portion of a polypeptide chain used to show the residue and peptide units

It is obvious that the backbone of the polypeptide serves as the framework supporting the side chains, and that the differences between side-chain groups are responsible for most topochemical features. However, it is also clear that the peptide group imparts its own direction and functionality on the many properties which polypeptides exhibit. In Sect.3.3.1 the backbone modifications are discussed. The peptide direction reversals (retro) accompanied by chain reversals (L- to D-analog conversion) are shown in Sects.3.3.2 and 3.3.3 to lead to inactive analogs. Chain bridging by disulfide bonds and carba-analogs are presented in Sect.3.3.4. Ordinary and abnormal amino acid substitutions (side chain modifications) are discussed in Sects.3.3.5 and 3.3.6.

3.3.1 Modifications to the Peptide Backbone

Modifications to the peptide backbone can be made by substitutions of atoms and groups of atoms at the α-carbon, the imino nitrogen, the carbonyl group, as well as intercalation of methylene groups within the chain at various positions. The effects that these modifications can have upon the conformation of the polypeptide will be described in the following sections, starting with a review of the conformational properties of the normal L-amino acids, continuing with D-residue substitutions, C^{α}-methyl additions, N-methyl, and depsipeptide modifications. A brief consideration will also be given to the "reduced" modifications at the carbonyl group, the intercalation of methylene groups, and the carbazic acid (α-Aza) derivatives.

a) *L-Amino Acids*

The model unit used most often to describe the allowed conformational space for amino acids is shown in Fig.3.2. The conventions used to define torsion (dihedral) angles are those adopted by IUPAC-IUB [3.23]. Briefly, a torsion angle ϕ about the bond B-C is defined by the sequence of atoms A-B-C-D. Looking down the B-C bond, the angle which the far bond, C-D, makes relative to the near bond, A-B, is the dihedral angle. If C-D is rotated clockwise from A-B the angle is positive; if counterclockwise, the angle is negative. Thus, a *cis* configuration would have $\phi = 0°$, while in the *trans* configuration, $\phi = 180°$. In the dipeptide shown in Fig.3.2, ϕ is defined as the torsion angle about the N-C^{α} bond, described specifically by the atoms

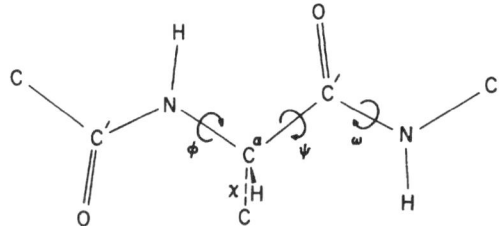

Fig. 3.2. A portion of a polypeptide chain used to show the dihedral angles ϕ, ψ, ω, and χ

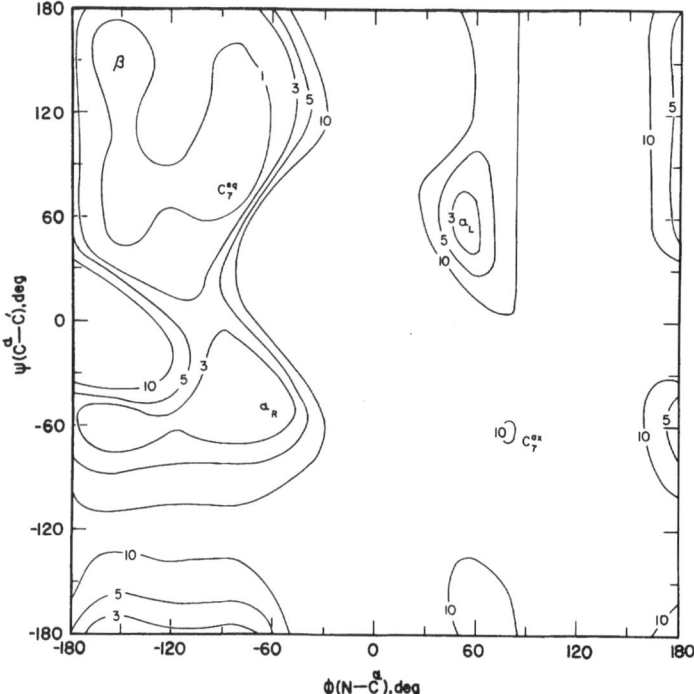

Fig. 3.3. A ϕ(N-C$^\alpha$), ψ(C$^\alpha$-C') conformational isoenergetic contour diagram for the molecule N-acetyl-N'-methyl-L-alanylamide. The energy is in kcal/mol

C'_{i-1}-N_i-C^α_i-C'_i, while ψ is defined as the torsion angle about the C^α-C' bond, described by the atoms N_i-C^α_i-C'_i-N_{i+1}. The peptide bond dihedral angle, ω, is defined by the atoms C^α_i-C'_i-N_{i+1}-C^α_{i+1}, while the side-chain dihedral angles, χ_i, are defined for each amino acid [3.23]. The variable dihedral angles ϕ, ψ, ω and χ are indicated in Fig. 3.2. Holding both peptide bonds in the *trans* (ω=180°) conformation, one can then consider the two dihedral angles ϕ and ψ as variables, by incrementing both dihedral angles through the range -180° to +180° by say 10° increments, calculating the total energy at each incremented point and smoothly connecting the isoenergetic points, one obtains a ϕ-ψ two-dimensional map, as shown for N-acetyl-N'-methyl-L-alanylamide in Fig.3.3.

The ϕ-ψ map shown in Fig.3.3 was obtained using the ECEPP program. Shown on the map are the symbols given to various standard conformations. For example, α_R (right-handed helix), α_L (left-handed helix), β (extended C_5 or beta structure), C^{eq}_7 (equatorial seven-membered hydrogen bonded ring), and C^{ax}_7 (axial seven-membered hydrogen bonded ring). The general features of the typical ϕ-ψ map shown are relatively insensitive to the calculating procedures used (i.e., semiempirical molecular orbital, ab initio, or other 'empirical' methods). The main differences between different computational methods arise in the magnitudes of the isoenergetic contours, and in some shifting differences in placements of the local energy minima. Observed ϕ-ψ

values obtained from X-ray structures of proteins, are found to fall within the 5 kcal/mol isoenergetic contour shown in Fig.3.3 [3.24,25].

Alanine, which has a methyl group as the side chain, could be expected to have a somewhat reduced allowed conformational space (i.e., within 5 kcal/mol of the global energy minimum) when compared to glycine, and indeed such is the case. As a general rule, one finds that those amino acids with side chains larger than a methyl group, and in particular, those with branched (either at the β or γ positions) substituents such as valine have much smaller regions of low-energy conformational space. However, the positions of minimum energy do not change dramatically from one amino acid to the next, only the size of the allowed low-energy regions. On the other hand, several amino acid residues do show fairly high populations of conformers in bridge regions (i.e., $\phi \sim 60°$ to $120°$ and $-60°$ to $-120°$, $\psi \sim 0$). These residues (i.e., glycine, serine, etc.) are those generally found in bends or chain reversal positions in proteins and are found in many small cyclic peptides. The bridge regions will be discussed later under bend conformations.

In general, one expects to find L-residues in the α_R, β, and C_7^{eq} regions of the $\phi-\psi$ map. However, a recent X-ray study of a small cyclic peptide [3.26] has shown clearly that the α_L position is also available to L-alanine. It should be noted that in this case there are (in the two molecules per unit cell) five L-alanine residues in the α_R and bridge region and only one L-alanine in the α_L region. This is a very clear-cut case which demonstrates the greater stability of the α_R over that of the α_L region, but indicates that under certain stabilizing conditions one could have an L-residue in the α_L configuration.

b) *D-Amino Acids*

A simple method for checking for L- or D-isomers is to proceed along the backbone from the C' atom, up to the C^α atom (as if going over a bridge) and on to the N atom. If the side chain at C^α points out to the left, it is the L-isomer, if it points to the right, it is the D-isomer.

The $\phi-\psi$ isoenergetic contour map of the N-acetyl-N'-methyl-D-alanylamide is shown in Fig.3.4. The symmetry with respect to the N-acetyl-N'-methyl-L-alanylamide $\phi-\psi$ map (Fig.3.3) is obvious, and all comments regarding the L-amino acid conformational preferences also pertain to the D-amino acids. Note that the preferred α-helix region now has positive ϕ and ψ values (ϕ=+,ψ=+), and C_7^{eq} and β structures have (ϕ=+, ψ=-). The notations, (+) and (-) will be used throughout this work to indicate a general region in $\phi-\psi$ space, favored by energetic criteria. Many X-ray structural studies have been carried out on peptides containing D-residues. Here we will give only a few structural results, choosing moderately large, flexible molecules in order to avoid the complications due to severe ring closure effects. For example, in cyclo-(Gly-Pro-Gly-D-Ala-Pro) [3.27,28], the experimental D-Ala ϕ and ψ values are (+134°,-69°). In the molecule cyclo-(4[Gly]-2[D-Ala]) the values of ϕ and ψ of

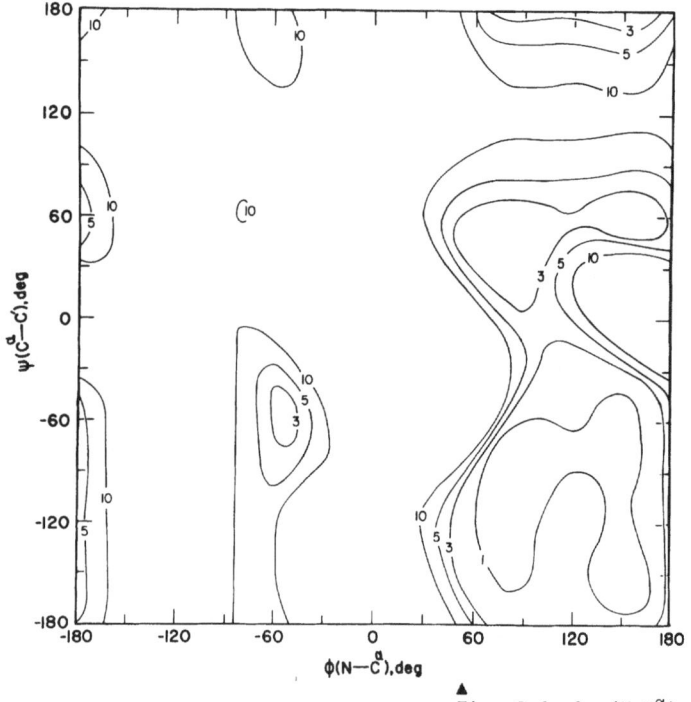

Fig. 3.4. A $\phi(N-C^\alpha)$, $\psi(C^\alpha-C')$ conformational isoenergetic contour diagram for the molecule N-acetyl-N'-methyl-D-alanylamide. The energy is in kcal/mol

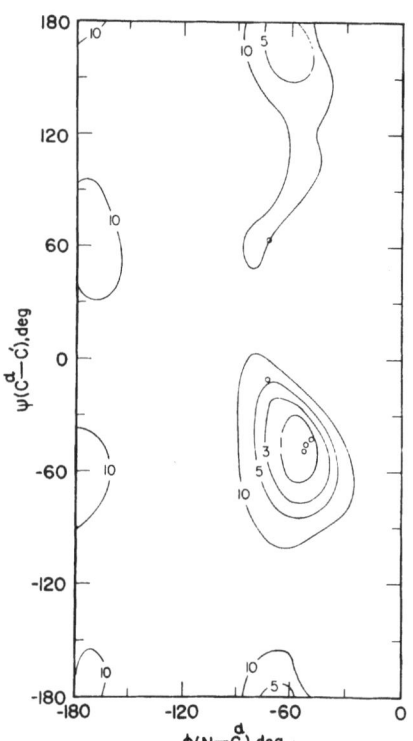

◄ Fig. 3.5. A $\phi(N-C^\alpha)$, $\psi(C^\alpha-C')$ conformational isoenergetic contour diagram for the molecule N-acetyl-N'-methyl-C$^\alpha$-methyl-alanylamide. Experimental values are shown as small circles. The energy is in kcal/mole. The region $\psi = 0°-180°$ is symmetric to that shown

Table 3.1. Experimental and calculated dihedral angles [deg.] for Aib residues in three polypeptides

	X ray		Calculated [3.35]	
	ϕ	ψ	ϕ	ψ
Aib^2 [3.32]	-52	-38	-50	-50
Aib^4 [3.32]	+48	+42	50	50
Aib^1 [3.33]	72	-64 (ω=162)		
Aib [3.34]	-56	-39 (ω_0=-173) (ω_1=-175)		

the two D-residues are found to be $(+66°,+15°)$ and $(+131°,-31°)$ [3.29]. In valino-mycin, two D-Val residues have $\phi \sim +60°$, $\psi \sim -135°$, while a third has $\phi \sim 108°$, $\psi \sim -69°$ [3.30]. These values are clearly within the calculated low-energy regions of the $\phi-\psi$ map shown in Fig.3.4 and are reasonably well symmetry-related to equivalent L-residues in similar cyclic structures.

c) C^α-Methyl Analogs

The $\phi-\psi$ map for N-acetyl-N'-methyl-C^α-methylalanylamide is shown in Fig.3.5. The α_R and α_L helical regions are the lowest-energy allowed conformations for C^α-methylated residues [3.31], with the C_7^{eq} and C_7^{ax} conformations being next lowest in energy. Structural evidence supporting the calculated regions of low-energy comes from the crystal structure study of Boc-L-Pro-Aib-L-Ala-Aib (Aib=aminoisobutyric acid). This molecule was found [3.32] to take up a conformation with the ϕ, ψ dihedral angles shown in Table 3.1. The agreement of other experimental values (dihydrochlamydocin [3.33] and N-acetyl-α-aminoisobutyric acid methylamide [3.34]) with calculated values [3.35] is also very good, as shown in Table 3.1.

The severely reduced allowed conformational space seen in Fig.3.5 places such strong limitations on a polypeptide conformation that its utility in polypeptide structure and drug design is particularly important.

Achiral C^α-methyl amino acids have been incorporated into a variety of polypep-tides, such as position one of ACTH 1-18 [3.36], positions three and five of angio-tensin (AII) [3.37,38], positions five and eight of bradykinin [3.39,40], and posi-tion two of methionine enkephalinamide [3.41]. The resulting biological activities were measured for these analogs and only [(αMe)Phe[8]]-AII retained ~100% activity. The other analogs had reduced but significant activity. One might interpret this re-sult to imply that the conformation at the Phe[8] position of the original AII molecule may be α-helical in nature, although the dihydrochlamydocin [3.33] structure suggests that the C_7 conformation must not be ruled out in these studies.

d) *N-Methyl Analogs*

The N-methylated residue and the $(i+1)^{th}$ N-methyl group relative to the ith residue
are shown in Fig.3.6. N-methylation of amide bonds is often undertaken in the hope
of protecting the peptide bond from attack by proteolytic enzymes. However, this
modification can profoundly alter the conformational preference as well as the potential hydrogen bonding capacity of the modified group. It is not surprising then,
that in some examples of N-methylation the peptide loses all activity, while modification in other positions on the same peptide enhances the activity. In this section we will attempt to describe, as simply as is possible, the steric and conformational affects of this modification.

Little work has been carried out by conformational energy calculations on complicated N-methyl analogs. However, several studies on model systems such as sarcosyl
[3.42,43] and N-methyl-L-alanine [3.44,45] yields a picture of the preferred conformations which should be applicable to most of the normal amino acids. The allowed

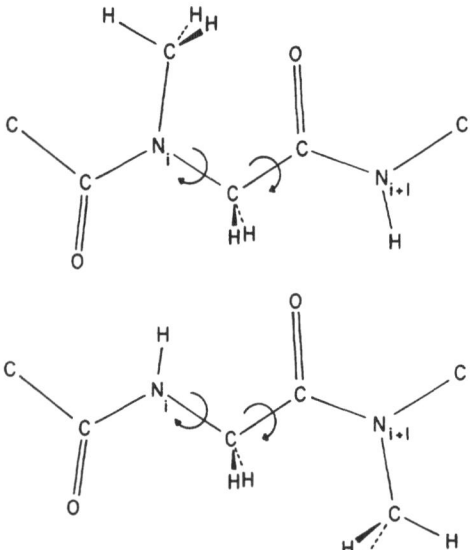

Fig. 3.6. A portion of a polypeptide
chain used to show the N_i-methyl and
N_{i+1}-methyl substitutions

Table 3.2. Calculated allowed conformations of sarcosyl residues [3.42]

	Dihedral angles [deg.]	
ϕ	ψ	ω
-70	160	180 (*trans* peptide)
-80	-180	180
-170	90	180
-150	80	0 (*cis* peptide)
-90	-110	0
-80	150	0

conformational space found for the sarcosyl group [3.42] is given in Table 3.2. The results of a similar calculation on sarcosyl-sarcosine [3.43] gave the minimum energy position as $\phi \sim +90°$, $\psi \sim -170°$, which is close to the C_7^{ax} position. However, a symmetry related C_7^{eq} position must also be of equal energy (i.e., $\phi \sim -90°$, $\psi \sim +170°$). It was found [3.44] that when the $(i+1)^{th}$ residue is N-methylated (see Fig.3.6), the minimum energy conformation for the ith residue occurred at $\phi \sim +90°$, $\psi \sim 180°$, with both α_R and α_L being conformations of low energy. In the case of N-methyl-L-alanine [3.44] the minimum energy position was found to occur at the α_L position (i.e., $\phi \sim 60°$, $\psi \sim 60°$). This result implies that a typical L-residue might go from an α_R to an α_L conformation upon N-methylation. Further complications occur to the ith alanine residue, when the $(i+1)^{th}$ residue is N-methylated (L-Ala succeeded by a *trans* peptide bond having an N-methyl group). In our calculations [3.35], the ith residue is shown to prefer a conformation in the region ($\phi \sim -140°$, $\psi \sim +80°$), with only a very small low-energy region in the α_L-helical regions. Recent calculations in this laboratory [3.35] gave the ϕ-ψ map for N-acetyl-N-methyl-N'-methyl-trans-alanylamide shown in Fig.3.7. The circles are ϕ-ψ values determined from available experimental

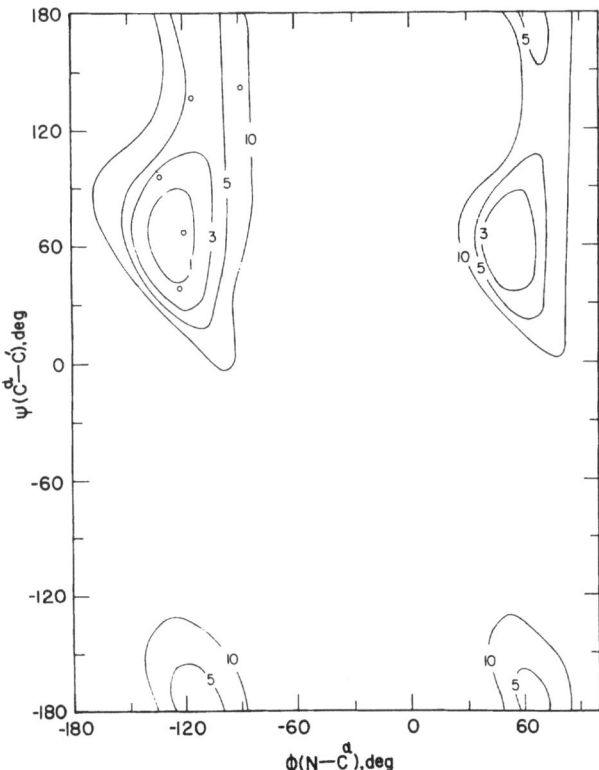

Fig. 3.7. A $\phi(N-C^\alpha)$, $\psi(C^\alpha-C')$ conformational isoenergetic contour diagram for the molecule N-acetyl-N'-methyl-N-methyl-trans-alanylamide. Experimental values are shown as small circles. The energy is in kcal/mol

Table 3.3. Experimental and calculated dihedral angles for N-methylated residues [deg]

	X ray		References	Calculated		References
Residue	ϕ	ψ		ϕ	ψ	
MeIle	-89	143	[3.48]	-110	170	[3.35]
MeLeu	-114	141	[3.48]	-110	170	[3.35]
cis-Sar	-94	180	[3.46]	-80	150	[3.42]
	94	180	[3.46]	80	-150	[3.42]
	-91	91	[3.47]			
	-87	108	[3.47]			
trans-Sar	120	-66	[3.46]	130	-70	[3.35]
	-120	66	[3.46]	-130	70	[3.35]
MeLeu	-128	99	[3.49]	-130	70	[3.35]
MeLeu	-121	38	[3.49]	-130	70	[3.35]

X-ray crystal studies [3.46-48]. We do not find that the α_L region is of lowest energy. Rather, the β region ($\phi \sim -130°$, $\psi \sim +70°$) is of lowest energy and is quite broad, leading one to surmise that both energetic and entropic contributions to the free energy enhance the β region preference.

Experimental studies have indicated that upon N-methylation, conformers with *cis* peptide bonds become populated in solution. Thus, one must always consider this possibility when looking at the conformational aspects of N-methylated analogs. The allowed sarcosyl conformers for $\omega = 0°$ are given in Table 3.2. Both sarcosyl studies [3.42,43] are in general agreement for the lowest energy conformation to be in the region $\phi \sim \pm 80$, $\psi \sim +150$. For the case in which the $(i+1)^{th}$ peptide bond is *cis* and N-methylated, the ith residue (alanine) again prefers the region $\phi \sim -150$, $\psi \sim +60 \rightarrow +180$, with only small probabilities for α_L conformations.

Good tests of these calculations are comparisons with various N-methylated molecules whose crystal structures have been solved by X-ray diffraction. Several sets of dihedral angles for the molecules 1) cyclotetrasarcosyl [3.46] and 2) cyclo-D-HyIv-L-MeIle-D-HyIv-L-MeLeu [3.48] are shown in Table 3.3 with comparative calculated allowed dihedral angles. As a general rule, one may be relatively safe in first taking the expanded C_7^{eq} structure ($\phi \sim -150°$, $\psi \sim 80°$) as being the most probable conformation for N-methylated L-amino acids and $\phi \sim +150°$, $\psi \sim -80°$ for the N-methylated D-amino acids. The second choice might be the α_L for L- and α_R for D-residue conformations. In any event, one would need to determine (say from NMR measurements) the configurations of the peptide bond (*cis* or *trans*) before making a final conformational judgment. One must also be wary of the effect on the ith residue of an N-methylated $(i+1)^{th}$ residue.

To conclude this section, the author knows of no case in which a N-methyl-C^{α}-methyl-L-amino acid has been used in analog studies. However, if one were to use this interesting combination, the result would be that only the region around $\phi \sim +60°$, $\psi \sim +60°$ would be populated with little deviation from these values.

The relationship between some physiologically active peptides and their N-methyl-ated analogs will be discussed in Sect.3.4.1b. Clearly, one must be aware that loss of physiological activity upon N-methylation may not simply be a result of a loss of a hydrogen bond (i.e., from removal of the N-hydrogen), but more probably may be the result of a severe conformational change brought about by this substitution.

e) *Depsipeptides*

The substitution of an oxygen atom for the N-H group of the peptide bond leads to ester linkages and molecules denoted as depsipeptide analogs, as is shown in Fig.3.8. The conformational properties of the depsipeptide analogs have been described elsewhere [3.50] and the calculated ϕ-ψ maps were shown to be very similar to normal dipeptides. In the case of the AA structure shown in Fig.3.8, the allowed low-energy space was considerably reduced in size, compared to the dipeptide (HA), but the α_R and β regions were still allowed and approximately in the same place on the ϕ-ψ map as found for normal dipeptides. Clearly, the lack of a N-H group eliminates the possible formation of the C_7^{eq} dipeptide conformation, since the hydrogen bonding ring can no longer be stabilized. Long-range hydrogen bonding would also be disallowed. However, in general, the depsipeptide looks remarkably similar to the normal dipeptide and must be considered to be isosteric to the peptide group. Analogs with depsipeptide linkages will be described in Sect.3.4.1c.

f) *Reduced Carbonyl, Intercalated Methylene (homo), and Carbazic Acid (α-Aza) Analogs*

The isosteric "reduced" analogs, carbazic acid (α-Aza) analogs and intercalated methylene groups, are shown in Fig.3.9. The author knowns of no studies on the conformational properties of these analogs. However, one can reasonably speculate on their conformational properties using standard model compounds as sources. First, the backbone flexibility of the "reduced" (i.e., $>C=O$ to $>CH_2$) as well as the inter-

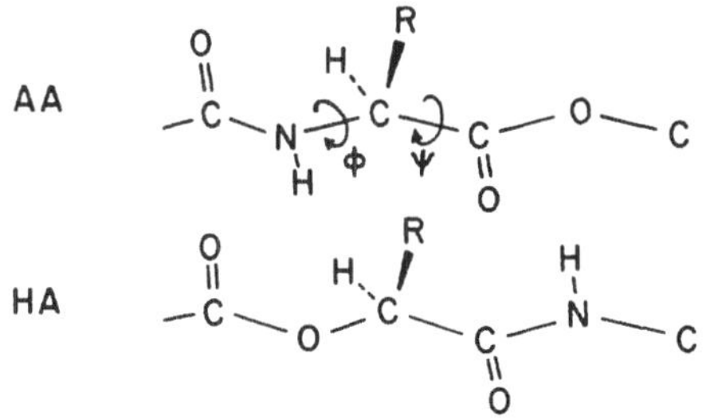

Fig. 3.8. A portion of a depsipeptide chain use to show the ester linkage

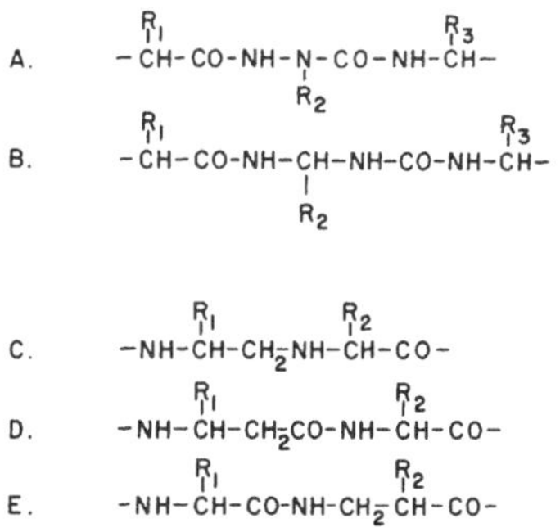

Fig. 3.9. (a) α-Aza derivative, (b) aza-α'-homo derivative, (c) reduced carbonyl derivative, (d)(e) intercalated methylene derivatives

calated methylene analog (homo) is greatly increased over that of a typical peptide. The constraint of near planarity of the peptide bond is completely removed in the re- duced case. The number of possible conformations which these molecules can take up is thus greatly increased, and one would anticipate that only for those conformers where the C^{α}-CH_2 and N-C^{α} bonds were *trans* to one another (for the reduced case) would the resulting structure closely resemble a native peptide. Clearly, the possi- bility of the C_7^{eq} structure would be energetically less favorable since the carbonyl group that was reduced to a methylene had been involved previously in a 7-membered hydrogen bonded ring.

The "carbazic" or α-Aza derivative is closely isosteric to a peptide group and retains the conformational preference dictated by the peptide bond. The geometry at the α-position (N^{α}) should mimic fairly closely either an L- or D-residue geometry, being somewhat more planar than the C^{α} geometry, and with bond angles closer to 120° than to $109^{\circ}28'$. The rotational barriers should not be significantly altered at the $\psi(N^{\alpha}$-C') bond.

The reduced susceptibility to enzymatic degradation is very important and these analogs must be considered as excellent choices for peptide bond protection, with retained conformational properties [3.51].

Molecules in which isosteric 'reduced' and carbazic acid have been introduced are: 1) renin inhibitors [3.51], 2) luteinizing hormone-releasing hormone (LHRH) [3.52-54], 3) angiotensin [3.55], and 4) gastrin C-terminal tetrapeptide amide [3.56]. The renin inhibitor analogs discussed are

a) $\quad$ H$_2$N-CH-CH$_2$-Leu-Val-Phe-OMe $\quad$, with CH$_2$-CHMe$_2$ side chain

b) $\quad$ H$_2$N-N-CO-Leu-Val-Phe-OMe $\quad$, with CH$_2$-CHMe$_2$ side chain

c) $\quad$ H$_2$N-N-CO-Leu-Val-NH-CH-CH$_2$OH $\quad$, with CH$_2$-CHMe$_2$ and CH$_2$Ph side chains

d) $\quad$ H$_2$N-CH-CH$_2$-NH-CH-CH$_2$-NH-CH-CH$_2$-NH-CH-CH$_2$OH $\quad$. with CH$_2$CHMe$_2$, CH$_2$CHMe$_2$, CHMe$_2$, and CH$_2$Ph side chains

All four peptides were found to be specific inhibitors of renin with potency equal or superior to that of the parent tetrapeptide ester, H-Leu-Leu-Val-Phe-OMe. The modified peptides were completely stable in the presence of leucineaminopeptidase, while under the same conditions the parent compound was rapidly hydrolyzed to the component amino acids [3.51].

In the LHRH example analogs [3.52], some antagonist activity was observed for [des-His[2]-(AzGly[10])]-LHRH, [des-His[2]-(D-Phe[6],AzGly[10])]-LHRH, and [D-Phe[2],D-Phe[6], AzGly[10]]-LHRH, but not for the AzGly[6] inhibitor analogs. In the case of LH agonists gastrin C-terminal tetrapeptide amide, and the sweeting agent L-aspartyl-L-phenyl-alanine methyl ester [3.56] and, incorporation of α-Aza analogs such as a) [AzGly[6]]-, [AzAla[6]]-, and [AzGly[10]]-LHRH were as active as LHRH; b) [AzPhe[4]]-, [AzAsp[3]]-, and [AzGly[4]]-amides of the gastrin C-terminal tetrapeptide were devoid of activity; and c) the dipeptide sweetener with the substitution [AzPhe[2]] was tasteless.

The angiotensin α'-homo-analogs were studied [3.55] in an attempt to elucidate some aspects of the hormone-receptor interaction. These Aza derivatives were discussed as being responsible for the spatial orientations of the side chains by the insertion of the extra NH-group into the peptide chain (Fig.3.9). The pressor activity of [Aza-α'-homo-L-Tyr[4]] AII, [Aza-α'-homo-D-Tyr[4]] AII, and [Aza-α'-homo-L-Val[3]] AII were all reduced to ~10-20% activity relative to AII. The published rationale [3.55] on the modified directionality of the side chains due to the inserted NH-group follows easily from model building studies of extended structures. However, one may argue that the urea group (Fig.3.9) could very easily lead to other unforseen conformational problems.

3.3.2 Chain Reversals and Bend Conformations

a) *LL, LD, and DL Bends*

Considerable evidence has accumulated suggesting that chain reversals or bends in small polypeptide hormones are conformational features important for physiological activity. Many workers have studied various aspects of bend conformations [3.57-68], and the following summary of bend types and conformational features will help put this unique structural feature in perspective for polypeptide drug design studies. Figure 3.10 shows one typical conformation found for LL type bends. In the figure shown, the backbone dihedral angles at i+1 and i+2 positions are, in terms of ϕ and ψ, (-,-)(-,+), where (-,-) implies the region $\phi \sim -50°$, $\psi \sim -50°$ (α_R helix) and (-,+) implies $\phi \sim -70°$, $\psi \sim$ (0 to +80°). A second bend type allowed for the LL sequence is defined as (-,-), (-,-) which has the α_R-α_R sequence, but that structure is far less energetically favorable than the one shown in Fig.3.10.

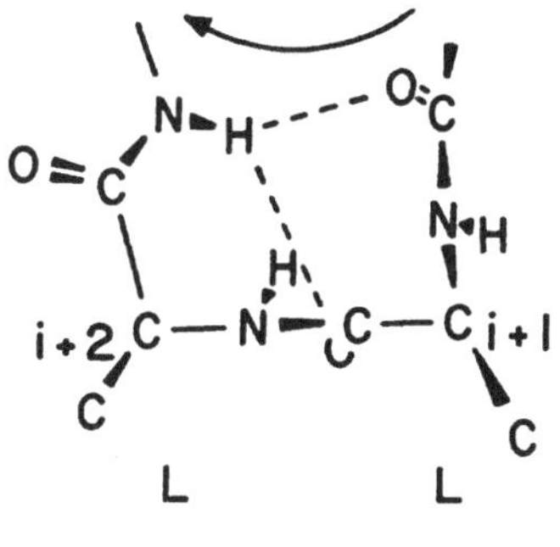

Fig. 3.10. The LL bend (-,-)(-,+). The dotted lines indicate favorable hydrogen bond type interactions. Both side-chain groups lie above the plane formed by the backbone atoms

Table 3.4. Observed and calculated conformational data

	Dihedral angles [deg.]				References
	ϕ_{i+1}	ψ_{i+1}	ϕ_{i+2}	ψ_{i+2}	
1. Bends in 10 proteins (mean values)					
Type I' and III'	51	107	98	-4	[3.61]
Type I and III	-53	-40	-74	-21	[3.61]
2. Structures					
Type I (LL)					
Theory	-50	-50	-110	40	[3.57]
	-64	-52	-83	85	[3.63]
Experimental					
L-Pro,L-Leu	-58	-33	-104	8	[3.69]
L-Pro,L-Leu	-65	-27	-105	8	[3.69]
L-Pro,L-Leu	-66	-29	-115	13	[3.69]
L-Ala,L-Phe(III)	-69	-13	-84	-6	[3.69]
L-Phe,L-Phe	-79	-13	-90	8	[3.69]
L-Pro,L-Lac(III)	-52	-22	-81	-11	[3.69]
Type I (GG)					
Gly, Gly	-69	-29	-94	8	[3.69]
Gly, Gly	-70	-15	-106	16	[3.69]
Type I' (DD) or (GD)					
Experimental					
D-Ala, D-Ala	66	15	131	-31	[3.69]
D-Hyv, D-Val	82	3	108	-69	[3.69]
Gly, D-Ala	74	12	134	-69	[3.27]
Type II (LG or LD)					
Theory	-60	100	60	40	[3.57]
	-87	82	52	58	[3.63]
Experimental					
L-Ser, Gly	-57	132	82	-1	[3.69]
L-Pro, D-Ala	-62	137	96	3	[3.69]
L-Pro, Gly	-52	126	74	12	[3.27]
L-Pro, D-Phe	-60	122	78	9	[3.28]
L-Leu, Gly	-61	128	72	–	[3.27]
L-Pro, D-Lac	-62	140	91	-8	[3.69]
L-Val, D-Hyv	-63	129	96	-3	[3.69]
L-Val, D-Hyv	-67	130	82	3	[3.69]
L-Pro, L-Ala	-59	136	66	14	[3.69]
L-Pro, D-Ala	-62	137	96	3	[3.69]
Type II' (DL or GL)					
Theory	71	-72	-69	-54	[3.63]
Experimental					
D-Val, L-Lac	63	-134	-74	-6	[3.69]
D-Val, L-Lac	60	-135	-98	14	[3.69]
D-Val, L-Lac	108	-69	-164	23	[3.69]
Gly, Pro	83	-134	-52	126	[3.27]
Type III (LL)					
Theory	-72	-60	-76	-57	[3.63]
Type III' (DD)					
Theory	54	46	56	47	[3.63]

When the $(i+2)^{th}$ residue is a glycine, then two more conformations become available. The first is $(-,+),(+,+)$ which would be equivalent to a $C_7^{eq}-\alpha_L$ set of dihedral angles, the second is $(-,+),(+,-)$ which represents a $C_7^{eq}-C_7^{ax}$ relationship. These latter two conformations are acceptable for LG, GG, or LD bend types, where G denotes a glycine residue and D denotes a D-residue. The LD bend types are included in the latter two forms since the D-residue would fit either a $(+,+)$ or a $(+,-)$ set of dihedral angles. Obviously, if a glycine occurring in a bend is replaced in the polypeptide by a D-residue, and the physiological activity remains, then one is fairly certain that the original conformation was made up of the two sets of dihedral angles noted above. The LD bend can also accommodate the sequences of LG, GD and GG.

The conformation for the sequences made up of DD or DL are exact inverses (or mirror images) of those for the sequences LL and LD, respectively, and have dihedral angle sets such as $(+,+)$, $(+,-)$ and $(+,+)$, $(+,+)$ for DD, and $(+,-)$, $(-,+)$ and $(+,-)$, $(-,-)$ for DL. Similarly as above, glycine can be at either corner position for DG, GG, or GL sequences. It should be noted that in LL or DD sequences the i+1 and i+2 positions, both C_{i+1}^{β} and C_{i+2}^{β} groups will be either above (LL) or below (DD) the plane defined by the backbone atoms (see Fig.3.10). In LD or DL sequences the same set of side-chain C atoms are on opposite sides of the backbone plane. These results will become increasingly important in polypeptide drug design and an example will be given in Sect.3.4.1c, showing use of this information.

An example of how one might interpret D- or L-substitution into a bend is given next. If a sequence such as A-LX-Gly-B, where A, B, and X are L-amino acids, is modified by substituting D-Y for Gly, and the analog retains activity or has enhanced activity, one would conclude that conformational sequences of the type: $(-,-)(+,+)$; or $(-,-)(+,-)$; or $(-,+)(+,+)$; or $(-,+)(+,-)$ [for (ϕ_{i+1},ψ_{i+1}) and (ϕ_{i+2},ψ_{i+2})] would be possible. However, to retain the bend at this site, only the combinations, $(-,+)$ $(+,-)$ and $(-,+)(+,+)$ are allowed since retaining $(-,-)$ in the $(i+1)^{th}$ position makes it impossible for the $(i+2)^{th}$ D-residue to fold back correctly (i.e., ϕ_{i+2} must be positive for a D-residue conformation). On the other hand, if the bend occurs in the sequence A-Gly-LX-B, then the GL sequences allowed are $(-,-)(-,-)$; $(-,-)(-,)$; $(+,-)(-,-)$; and $(+,-)(-,+)$. When a D-residue is substituted for glycine the remaining allowed combinations are $(+,-)(-,-)$ and $(+,-)(-,+)$.

In order to evaluate the conformational conditions described above, the data of Table 3.4 are given to present a variety of conformational data from X-ray structural studies and energy calculations. If each group or type is averaged over observed and calculated structures, the general rules for bend conformations would look as follows

	ϕ_{i+1}	ψ_{i+1}	ϕ_{i+2}	ψ_{i+2}
Type I (LL) $(-,-)(-,+)$	$-80 \rightarrow -50$	$-50 \rightarrow -10$	$-100 \rightarrow -70$	$85 \rightarrow 8$

	ϕ_{i+1}	ψ_{i+1}	ϕ_{i+2}	ψ_{i+2}
Type I' (DD) (+,+)(+,-)	70 → 50	110 → 15	130 → 75	-70 → -4
Type II (LG or LD) (-,+)(+,+)	-90 → -50	140 → 80	100 → 50	60 → 0
Type II' (DL or GL) (+,-)(-,-)	70 → 50	-140 → -70	-100 → -70	0 → -50
Type III (LL) (-,-)(-,-)	-70 → -50	-60 → -10	-90 → -70	-60 → -5
Type III' (DD) (+,+)(+,+)	70 → 50	60 → 40	60 → 40	50 → 0

The rationale for using calculated values as well as experimental values in the above averages is to include some weight for free molecule conformations, since many of the experimental values are in small cyclic molecules, where the distortions from ring closure are generally more severe than for free linear molecules. Further, the effects from crystal packing forces have not yet been fully examined for these bend conformations, and the observed flattening effects (i.e., $\psi \sim 0°$) could arise from crystal packing effects. Further work will be required to examine this problem.

The consequences of replacing an L-residue (in an all-L-polypeptide) with a D-residue are much more complex and require significant backbone conformational changes to be made in order to retain nearly equivalent orientations of side chains. One way this change can be made will be discussed for the L- to D-Trp[8] modification in the growth hormone releasing inhibitor, somatostatin (see Case IV in Sect.3.4.1a). Each of the types of bends shown with their respective sets of dihedral angles given in Table 3.4 form bends, and as can be seen, proline is a common amino acid in many structures in which bends are observed. Further it should be noted that most of the ψ_{i+2} dihedral angles have small positive or negative values. This is most probably a result of forming a good hydrogen bonding interaction between the N-H of the $(i+2)^{th}$ residue to the C=O of the ith residue, but may also be a result of crystal packing forces as described earlier.

It has been suggested [3.70] that bends are favored in Pro-X sequences because ϕ_{Pro} is fixed by the geometry of the pyrrolidine ring in a conformation frequently found in bend structures. However, bends are not usually favored in X-Pro sequences because interactions between X residues and the pyrrolidine ring restrict the X residue to conformations which are not usually found in bends. The use of proline and proline homologs as probes of conformation will be examined in Sect.3.3.6.

b) *Depsipeptide Bend Structures*

It is important to note that depsipeptide modifications to the backbone also have the capability of taking up bend conformations. Conformational energy calculations and structural data have shown that the LL and LD sequences shown in Fig.3.11 can

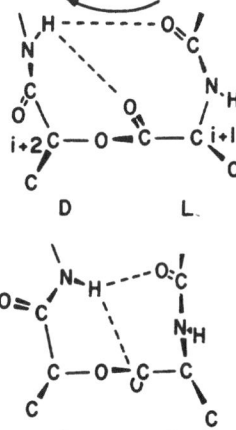

D   L.

L   L

Fig. 3.11. The LL and LD bends for the depsipeptide in the i+1 position. The dotted lines indicate favorable hydrogen bond type interactions

Table 3.5. Experimental dihedral angles [deg.] for depsipeptides [3.71,72]

	ϕ_{i+1}	ψ_{i+1}	ϕ_{i+2}	ψ_{i+2}
L-Pro L-Lac	-55	-22	-81	-11
L-Pro D-Lac	-62	140	91	-8

take up bend structures similar to their peptide analog equivalents [3.71]. From Fig.3.11 one can see that the only change when going from LL to LD or DD to DL is the flipping of the ester linkage which can be designated as going from LL (-,-) (-,$\bar{+}$) to LD (-,+)(+,$\bar{+}$) or from DD (+,+)(+,+) to DL (+,-)(-,$\bar{+}$). Experimental evidence is presented for depsipeptide structures in Table 3.5. Note that these two conformations are remarkably similar to those found for LL and LD bends in Table 3.4.

3.3.3 Direction Reversal: Retro, Retroenantiomers, and Retro-all-D-peptides

One interesting approach to the study of topochemical features of hormones and hormone analogs is to observe the effect of direction reversal of the peptide bonds and subsequent enantiomeric configuration change in the side chains. Simply reversing the direction of the amide bonds results in a "retro"-peptide, while retro with simultaneous L- to D-conversion results in retro-all-D; or if end-group effects are correctly treated, to a retro-enantiomer of the parent polypeptide.

Recent studies [3.73,74] have enumerated many examples of retro-enantiomers of both cyclic and linear peptide hormones and enzyme substrates. As a general rule, the probability for activity of retro-enantiomers is quite high for cyclic peptides and very low for linear peptides. In both cyclic and linear cases, a comparison of molecular models of the native and retro-enantiomer show close correspondence in molecular architecture, except for the reversed direction of the peptide bonds (Fig.3.12).

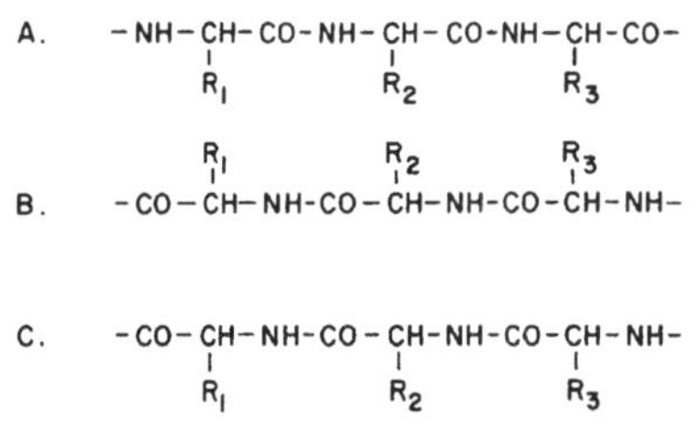

A. $-NH-CH-CO-NH-CH-CO-NH-CH-CO-$
           | | |
           R_1 R_2 R_3

B. $-CO-CH-NH-CO-CH-NH-CO-CH-NH-$
         | | |
         R_1 R_2 R_3

C. $-CO-CH-NH-CO-CH-NH-CO-CH-NH-$
         | | |
         R_1 R_2 R_3

Fig. 3.12a-c. A portion of a polypeptide chain used to show (a) normal, (b) retro-, and (c) retroenantio-configuration

Table 3.6. Retroenantiomers and retro-all-D-peptides and their biological activity

Analog	Activity
Retro-enantio-bradykinin [3.74]	Inactive
Retro-enantio-α-melanotropin(5-9)-pentapeptide [3.74]	Inactive
Retro-all-D-tuftsin [3.74]	Inactive
Retro-enantio-desamino-gastrin-C-terminal tetrapeptide amide [3.74]	Inactive
Retro-enantio-α-amino malonyl-D,L-phenylalanine [3.76]	Not sweet
Retro-des-Gly^{10}-[Pro^9-NEt]-LHRH [3.77]	Inactive
Retro-[L-Phe^2]-LHRH [3.76]	Inactive
Retro-enantio-[Suc^1]-angiotensin II [3.78,79]	Inactive
Retro-enantio-[Suc^1,Ala^7]-angiotensin II 24% of [Suc^1,Ala^7]-AII [3.78,79]	Active
Retro-enantio-somatostatin [3.80]	Inactive
Retro-all-D-Leu-enkephalin [3.81]	Inactive

Although the reversed peptide bond and subsequent changes in hydrogen bonding have been held responsible for the lack of biological activity of the linear peptides [3.74], recently it was shown [3.75] by conformational energy calculations that one can expect large and significant conformational differences between the low-energy conformations of the parent and retro-all-D-molecules. It seems very logical to blame the lack of biological activity on lack of correct conformation since some cyclic peptides whose conformations are held quite rigidly by the cyclic structure show high biological activity. Many linear retro-all-D-molecules and retro-enantiomers of biologically active polypeptides have recently been synthesized and tested for physiological activity, and the results are shown in Table 3.6. The retro-enantio-[Suc^1,Ala^7]-angiotensin II analog [3.78,79] seems to be the first linear peptide which retained any significant biological activity. Although these retro-molecules were inactive as a rule, it will be shown in Sect.3.4.1c that it is possible to incorporate the retro- and LD-combinations into molecules and retain or enhance activity or change agonist activity into antagonist activity. The solution is to design the L- and D-combinations so that the resulting low-energy conformation is similar in stereoequivalent ways to the active molecules. Thus, one must be very clever in the use of structural design in order to obtain physiologically active

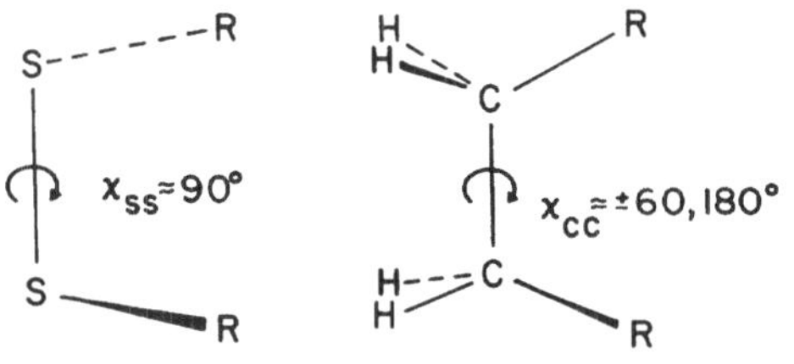

Fig. 3.13. View of the disulfide bond and the stereoequivalent carba-analog (both sulfur atoms substituted by methylene groups)

structures. Simply synthesizing the retro-all-D or retro-enantio molecule will not be a particularly profitable venture in most cases of linear molecules.

3.3.4 Bridges: Disulfide Bonds and Carba-Analogs

In many cases where the disulfide bridge is found, one may wish to establish whether the role of the disulfide bond in the biologically active polypeptide is simply structural (i.e., structural rigidity, entropy loss, etc.) or has some essential chemical property connected to it. Several analogs have been examined which closely resemble the geometry of the disulfide bridge, but which lack the reactive disulfide bond. An analog in which *one* sulfur is replaced by a methylene ($-CH_2$) group would meet the two requirements of structural similarity and be somewhat isosteric. Analogs of this type (carba-) were examined in oxytocin some years ago [3.82]. The oxytocin analog examined [3.82] was active and led the authors to conclude that the disulfide bond in oxytocin was a structural, not chemical, requirement for activity. Structural similarities of the disulfide bridge with both sulfur atoms replaced by forming the carba-analogs are shown in Fig.3.13. The barrier to rotation about the disulfide bond is $\sim$14 kcal/mol with minima at $\chi_{xx} \approx \pm 90°$, while in the C-C bond of the carba-analog, the barrier is only $\sim$3 kcal/mol. Thus, the carba-analog can very easily take up a χ value near $\pm 90°$, even though the more favorable conformations would be $\chi \approx \pm 60°$, $180°$, which are the typical hydrocarbon gauche and *trans* conformations.

Recent confirmation of the stereo equivalents between the carba- and disulfide bonds comes from a NMR comparison of the model compounds cyclo-L-cystathionine and cyclo-L-cystine [3.83]. Both cyclo compounds exhibited the same configuration.

Recently, several interesting carba-analogs of the growth hormone inhibitor, somatostatin, have been tested for physiological activity [3.84-88]. In nearly every case the activities were equivalent to the native hormone, with only subtle differences in relative activity between inhibition of growth hormone, glucagon or insulin.

3.3.5 Modifications of Side Chains

By far the most popular approach to the modification of peptide structure has been to replace the naturally occurring constituent amino acids by other naturally occur-

ring amino acids [3.82]. The procedure generally used is to replace each individual
amino acid in the sequence by alanine, one at a time. In this way, the correlation
of each site with the molecular activity is obtained, and if a particular change
exhibits a dramatic difference in activity, then that site is generally said to be
taking part in the "active" region or is part of the "binding site", etc. A second
approach often used is to substitute a homologous amino acid for the native one. In
this approach, a basic side chain is replaced by a second basic group (i.e., Arg
for Lys, etc.) or a nonpolar side chain for a second nonpolar one (i.e., Leu for
Ile or Met) or one ring-bearing side chain for another (i.e., Phe for Trp or Tyr).
The variety of possibilities is truly overwhelming and can easily lead to so many
possible analogs that the costs of synthesis and testing become prohibitively high.
To reduce these numerical problems to a reasonable level one must use some rather
stringent guidelines. From a conformational standpoint, only proline of all the natural amino acids is truly useful in testing particular conformational constraints.
For example, if an L-residue normally is found to require a conformation in the
range of $\phi \sim -50^\circ$ to -100° (this involves both α_R and C_7^{eq} conformations), then substitution of proline ($\phi \sim -75^\circ$) could lead to a similar structure. The assumption
one makes is that the polarity of the residue is such that the "active site" is not
modified. Thus, a positive activity result upon Pro substitution gives one significant structural information, but a negative result could be due to reasons other
than conformational change.

3.3.6 Abnormal Side-Chain Modifications

It is recognized that normal amino-acid side chains can influence the preferred
backbone conformation, however, it is most generally true for side-chain environ-
ments in the interior of proteins, where other steric factors may impose rather severe restrictions on allowed conformations. In the case of most small polypeptides
it is generally found from NMR and computed conformations that several side-chain
conformations may be equally probable. For this reason we will, in this chapter,
discuss only those conformational effects which *must* have significant conformational
effects on the backbone conformation. Those modifications which *might* have significant conformational effects are not included here.

a) α, β-*Dehydro Analogs*

The unusual α,β-unsaturated amino acids (see Fig.3.14) have been found in the hetero-
detic pentacyclic peptides, nison [3.89,90] and subtilin [3.91], in the forms of
α,β-dehydroalanine (Dha) and α,β-dehydrobutyrine (Dhb). The α,β-dehydrophenylalanine
(Phe[(Z)Δ]) residue has been found in tentoxin [3.92], dehydroserine ureide in viomycin [3.93], and dehydrotryptophan in telomycin [3.94]. NMR structural studies indicate that the RCH $= C^\alpha - C' = 0$ unit is conjugated and nearly planar [3.92]. The most
probable configuration is that in which the carbonyl bond is *cis* to the α,β-unsatur-

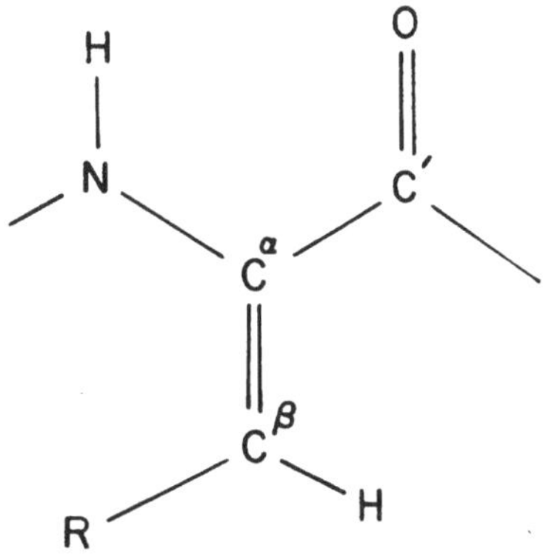

Fig. 3.14. A view of the α,β-dehydro-residue

ated bond and nearly coplanar (i.e., $\psi \sim \pm 30°$). A similar structure around the serine ureide residue in viomycin was shown by X-ray analysis [3.93]. Thus, taking the dihedral values suggested for L-Pro[1]-tentoxin and tentoxin [3.92] ($\phi \sim -90°$, $\psi \sim -20°$) and the projected values for the three conformers suggested for D-MeAla[1]-tentoxin [($\phi \sim -90°$, $\psi \sim 160°$); ($\phi \sim -80°$, $\psi \sim -30°$); and ($\phi \sim +90°$, $\psi \sim -20°$)] one can suggest that $\phi \sim \pm 90°$, $\psi \sim -20°$ or $+160°$ will be generally in the correct region of conformational space. To date, only one conformational study has been carried out on one of these interesting abnormal amino acids [3.92]. However, their use as a probe of the conformation of polypeptides is obvious, and future work will no doubt provide further insight into their conformational effects [3.95,96].

b) 3,4-Dehydroproline, Azetidine, and Thiazolidine

The introduction of abnormal residues such as 3,4-dehydroproline [3.97-99], azetidine [3.100-103], and thiazolidine (Thz) [3.104-106] (see Fig.3.15) into normal proline positions in a polypeptide chain should be useful as tests for steric and structural properties at that site. In particular, these residues should retain the β-bend characteristics of proline, while at the same time allowing different regions of the ψ dihedral angle to be favorable or unfavorable. It is clear that incorporation of 3,4-dehydroproline in place of proline can make a fundamental difference in physiological activity [3.97-99]. However, the question remains whether the activity response is a result of π-π bonding, as suggested [3.97,98], or whether this could be another case in which increased flexibility (larger regions of allowed conformational space) allowed a more favorable structure at the receptor. Clearly, by making the double bond, the conformational freedom for rotation about the ψ dihedral angle has been increased considerably [3.107a]. A second effect, as yet undetermined, is the conformational change the 3,4-dehydroproline makes in the φ dihedral angle, relative to $\psi \sim -75°$ in proline. A recent crystal structure study gave φ values of $-60°$ and $-64°$ [3.107b], which are $\sim 15-20°$ less than the φ values found in L-proline studies. One final point on the 3,4-dehydroproline analog is that it probably imparts increased resistance to enzymatic degradation during passage through the pulmonary

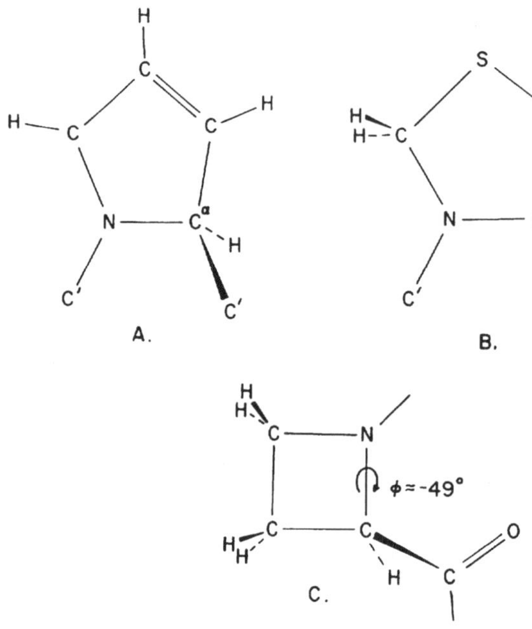

Fig. 3.15a-c. A view of (a) 3,4-de-hydroproline, (b) thiazolidine, and (c) azetidine residues

circulation system [3.99]. This result was noted for the analogs [L-Δ^3Pro^2]- and [L-Δ^3Pro^3]-bradykinins [3.99], which were found to be as potent as native brady-kinins [3.99].

In the case of azetidine (see Fig.3.15), the X-ray structure is available for the square-ring [3.101] and the projected ϕ dihedral angle of $\sim$-49° is considerably smaller than that found for proline ($\phi \sim$-75°). The possibility for an increase in the allowed range of the ψ dihedral angle (over that of proline) was pointed out in a study on prolyl-azetidine-2-carbonyl-proline [3.100].

The case of the thiazolidine residue is somewhat better known. The X-ray diffrac-tion study has detailed the geometry of the ring, and when compared to an equivalent proline model compound, the geometries and conformations are very similar [3.104, 105]. In thiazolidine, $\phi_{thz} \sim$-85° was close to the observed proline value $\phi_{pro} \sim$-80°, while $\psi_{thz} \sim$-9° is comparable to $\psi_{pro} \sim$-14°. The ring dihedral angles were also very similar. Clearly, the thiazolidine ring is closely isosteric to proline and should exhibit very similar conformational properties. Substitution of thiazolidine into thyotropin releasing factor (TRH) in place of proline resulted in $\sim$16% activity re-lative to TRH [3.104]. The enkephalin analogs, [D-Thr^2,Thz^5]-EK-NH$_2$ and [D-Met^2, Thz^5]-EK-NH$_2$, were also found to be very active [3.106].

3.4 Application of Conformational Information to Drug Design

The conceptual goals or aims of peptide drug design are not easily stated. They may
arise from the need for a wide variety of desirable features, such as: 1) increased
potency, 2) increased selectivity (even at the expense of potency), 3) a longer time
course of action, 4) more potent inhibitors or antagonists, 5) better lipophilic
properties (i.e., ability to partition from an aqueous solution such as the blood
through a lipid barrier such as the blood-brain barrier), 6) better resistance to
enzymatic attack (but perhaps not complete resistance), or 7) better availability
after "first pass" effects. Finally, neither the parent molecule, nor its products
must be toxic. As if these problems were not enough, one must often be concerned
about how the drug tastes, its smell, whether it is painful at the injection site,
and finally its formulation for oral application (liquid or tablet?). For new potent
analogs one must ask, whether they have any practical application or scientific
usefulness. All of the above features are important and must ultimately receive some
attention in the design problem. However, it would be beyond the scope of this chapter to describe recent examples of attacks on each of these features and so we will
concentrate only on some aspects of current interest in which the peptide structure
is important.

It should be obvious at this juncture that the author feels that if one had available a working model or conformation of the molecule of interest, the systematic design of biologically active analogs could proceed along rational structural guidelines. It is usually preferable to have an experimentally determined structure,
rather than one from conformational energy calculations. However, there are many
problems associated with the determination of experimental conformations in solution,
thus, the calculated structures are often the only ones available. Calculated structures must be used very carefully, as pointed out in the introduction, but it seems
important to stress here that it could very well be more profitable to design analogs
to *provide structural data* early in the game than to use interesting, but structur-
ally unrelated analogs when starting a program to prove "binding sites", "active
sites", etc. It follows that once the active structure is known, the design of new
agonists or antagonists will proceed much more rapidly, and more economically. One
further word of warning may be in order here: one must realize that a modification
designed to enhance resistance to proteolytic attack may very well cause a conformational change which destroys the desired activity, thus negating the change. This
will be seen to be particularly important in the cases of L- to D-residue substitution.

3.4.1 Example Studies: Backbone Substitutions and Conformational Effects

In this section we will attempt to examine some recent analog data to show how
structural modifications, which are closely coupled to conformational effects, can

affect activity. The examples chosen are not meant in any sense to be a comprehensive listing of the complete series of analogs for any single peptide. Further, in some cases the interpretations as to possible conformational effects will be those of *this* author, and not of the authors of record.

a) *D-Amino Acid Substitution*

It was pointed out earlier that the side chains of enantiomeric amino acids are fully identical functionally, but that a peptide chain replacement of an L- by a D-amino acid residue may lead to profound steric changes. Although at first sight it would appear that such L- to D-changes at the two ends of a linear peptide chain would least affect the overall chemistry of the molecule and might only displace the terminal carboxyl, carboximide, or amino groups, there are recent examples where end-group modifications exhibit interesting and significant results. Examples of these groups are described in this section (Case I,II).

The consequences of L- to D-changes *within* the polypeptide chain will vary from case to case and are difficult to predict. Nevertheless, such changes have proven to be popular. Several such studies on oxytocin, angiotensin, melanotropin, and corticotropin fragments have been reviewed [3.82], and we will not extend our discussion to the more recent analogs of these molecules. However, many examples of L- to D-modifications may be given from the recently discovered peptide hormones and the opiate-acting peptides (enkephalins). To stay within the conformational framework of this chapter, we will only include those modifications which contain significant structural information.

Case I: C-terminal of Somatostatin. Somatostatin (SS) is a cyclic tetradecapeptide with the sequence H-Ala-Gly-Cys-Lys-Asn-Phe-Phe-Trp-Lys-Thr-Phe-Thr-Ser-Cys-COOH with a disulfide bridge between cysteine residues. Many analogs of SS have recently been prepared, but of interest here is the D-Cys^{14} analog which takes part in the disulfide linkage. It has been found [3.108] that this analog is responsible for enhancing the separation of functions that SS exhibits. For example, the inhibition of growth hormone (GH) is enhanced over that of normal SS (240% relative to 100% for SS), while insulin inhibition is remarkably reduced (10% compared to 100% for SS), and glucagon inhibition is unaffected by this change. Examination of a calculated structure [3.6] (Fig.3.16) shows that the major effect that this analog has upon the structure is to remove the C-terminal carboxyl group from an exposed (external) position to a more buried (internal) position, with little or no effect on the disulfide bond. This effect is very subtle and may be no more than a change in the ability of the molecule to reach one receptor (GH) more easily than another (insulin). This conclusion is purely speculation on the part of this author.

Case II: N-terminal of Luteinizing Hormone-Releasing Hormone (LHRH). Luteinizing hormone-releasing hormone (LHRH) is a decapeptide with the sequence: pGlu-His-Trp-Ser-Tyr-Gly-Leu-Arg-Pro-Gly-NH$_2$. Many analogs have been prepared in recent years in

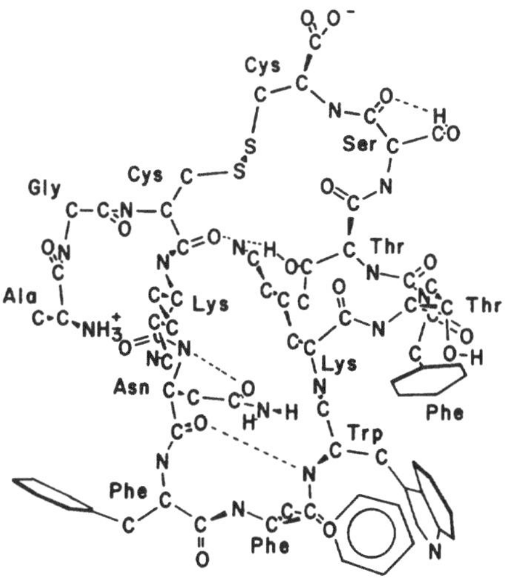

Fig. 3.16. A view of a conformer (3D) of somatostatin. Dotted lines indicate favorable hydrogen bond interactions. For clarity, some hydrogen atoms have been omitted

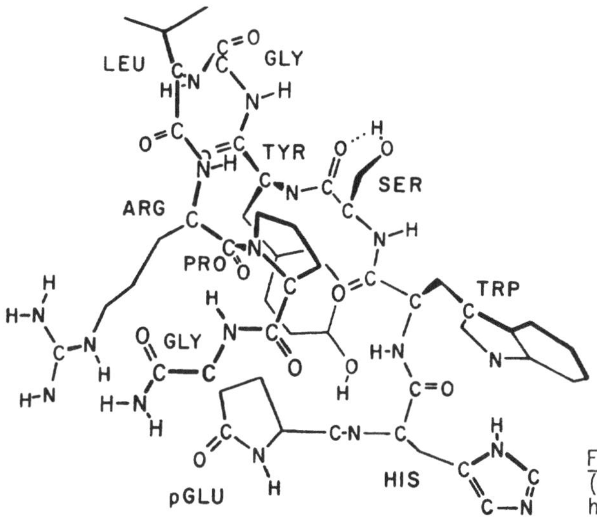

Fig. 3.17. A view of a conformer (CC) of LH-RH. For clarity, some hydrogen atoms have been omitted

an attempt to produce more active agonists as well as antagonists. The [D-pGlu^1]-LHRH analog was found [3.109,110] to have reduced agonistic activity (~10%) relative to native LHRH (100%). A calculated structure of LHRH is shown in Fig.3.17. One might expect that by substituting D-pGlu^1 for L-pGlu^1 the major result would be to turn the *cis* peptide of the pGlu ring into the molecule, rather than pointing out as shown in Fig.3.17. This is one possible explanation for the drop in activity and can certainly be taken to mean that some part of the pGlu ring is involved in the active site of the molecule. We will see later (Case V) that this residue exerts an important influence in antagonist activity.

Case III: Substitution for Glycine in Enkephalins (EK) and LHRH. Recent examples of
probing the structure-activity response of a glycine residue are found in studies
on methionine and leucine enkephalins [3.111-116]. These small five residue peptides
have the sequence: H-Tyr-Gly-Gly-Phe-Met-COOH and H-Tyr-Gly-Gly-Phe-Leu-COOH. They
are endogenous to mammals and are thought to be the molecules responsible for the
occurrence of brain receptors which recognize the opiates such as morphine [3.117].
A very early analog which was found to be ~5-8 times more potent than the natural
EK molecules was [D-Ala^2]-EK [3.111-116]. Since some other D-X^2 analogs also enhance
analgesic activity [3.111,115,116], we are fairly safe in assuming that the conformation around the *two* position is favorable for the dihedral angle values of (ϕ=+;
ψ=±). It should also be noted that the L-Ala^2 analog loses most of its activity,
which further supports this conclusion. Substitutions at the Gly^3 position are nearly always destructive of activity. Neither L- or D-residues at Gly^3 have been found
which substantially enhance activity. This is a remarkable situation and is the only
case known to this author in which a glycine appears not to be replaceable by any
other *L- or D-residue*. Further work on this question is in progress in this laboratory [3.118].

    A second important example of a glycine being replaced by a D-residue occurs at
the six position (normally Gly) of LHRH (see Case II above). In this molecule it was
found [3.119,120] that [D-Ala^6]-LHRH substantially enhanced the agonist activity of
LHRH by 3-5 times. Other [D-X^6]-LHRH analogs also where shown to be particularly
active [3.121-123], and further conformational energy studies [3.1,2] showed that
this substitution was favorable for stabilizing the molecular conformation. This
substitution also imparted some degree of stability against proteolytic attack
[3.122]. [L-X^6]-LHRH analogs were all found to decrease the agonist activity dramatically.

    Further evidence for the conformation of LHRH around the six position will be
discussed in Sect.3.4.1b (Case VI).

Case IV: L- to D-Conversion in Somatostatin. The systematic substitution of each
L-amino acid in somatostatin with its D-isomer has been carried out [3.123,124].
Looking at just the growth hormone inhibition from in vitro studies with SS taken
as 100%, we see from Table 3.7 the following:

1) As a general rule, the D-substitutions at the N- and C-terminal ends of the molecule retain significant activity (see Case I).
2) The middle residues (6-10) are, with the exception of the D-Trp^8 analog, relatively inactive.
3) D-Trp^8 is by far the most active analog, by simple single D-substitution.

    We might ask, what are the conformational criteria that must be satisfied when
an L-residue is substituted by its D-isomer in order to retain some structural homo-
logy with the active native conformer? The answer to this is somewhat complex and
requires subtle changes in the backbone dihedral angles of the residues on each side,

Table 3.7. Activities of somatostatin analogs on growth hormone inhibition

Analog	% Potency [3.123]
[D-Ala^1]-SS	200
[D-Ala^2]-SS*	55
[D-Cys^3]-SS	50
[D-Lys^4]-SS	22
[D-Asn^5]-SS	0.6
[D-Phe^6]-SS	5
[D-Phe^7]-SS	1
[D-Trp^8]-SS	800
[D-Lys^9]-SS	1
[D-Thr^{10}]-SS	0.5
[D-Phe^{11}]-SS	10
[D-Thr^{12}]-SS	20
[D-Ser^{13}]-SS	10
[D-Cys^{14}]-SS	270

* Gly is the "normal" residue on position 2.

particularly on the N-terminal side of the substitution site. An example may help clarify this point. Assume the sequence X-Y-Z with the sets of ϕ and ψ dihedral angles: X(-,-), Y(-,+), Z(-,+). Now substitute D-Y for L-Y. In order to move the side chains of X and Z by the least amount, the peptide bond on the N-terminal side of D-Y must be flipped by 180°. This rotation results in X(-,+), D-Y(+,+), Z(-,+) and the X and Z side-chain positions in space are only minimally changed. However, suppose that in the original sequence the X and Y side chains were both pointing down from the plane defined by the backbone atoms; in the new structure the X side chain still points down, but the D-Y side chain points upward. The rest of the mole-cule is relatively unchanged. A second possibility exists for L- to D-conversion when the L-residue to be modified is in the fully extended conformation (i.e., $\phi \sim -160^\circ$; $\psi \sim 150^\circ$) in its native structure. In this case, no peptide group rotation need be carried out, since one can easily move to the extended conformation for the D-residue (i.e., $\phi \sim +160^\circ$; $\psi \sim -150^\circ$) without large distortions in the dihedral angles of the neighboring residues. This modification suggests that one cannot say with certainty that a favorable L- to D-conversion means that the original L-residue was part of a bend configuration.

Returning to the somatostatin example, we can now understand somewhat the results of these L- to D-changes on the activity. For example, changes at those positions that have the least effect on the active site should show the least effect on ac-tivity, while those near the active site should be most damaging (or enhancing) to the activity. From this one anticipates (Table 3.7) that the stereo-positions of the side chains of residues 5 through 10 are very sensitive to subtle changes in

those residues. An alternative explanation could be that those residues are important for the conformational integrity of the molecule and that by flipping the peptide bond, this integrity was broken. This author tends to lean toward the former
explanation, but much conformational work along these lines is needed.

We are left to explain the greatly enhanced activity of the D-Trp^8 analog. As
noted previously, this change would place the Phe^7 side-chain ring and D-Trp^8 side-
chain ring with one up and the other down, assuming a bend conformation as shown in
Fig.3.16. The result of this change is to allow the two rings to stack one above
the other such that a ring-to-ring gap of ~7 Å results. Other evidence [3.125] to
be described for the GH agonists suggest that ring stacking is the key identifying
feature at the GH receptor. We believe that this stacking allows a specific receptor
group to intercalate between, say, the Phe^7 and D-Trp^8 rings and form a stable complex of aromatic rings. This complex will also be examined as an explanation for
the activity of GH releasing factors [3.125]. Although we have noted Phe^7 as the
second ring, it is entirely possible that Phe^6 stacks with the D-Trp^8 ring, while
Phe^7 and Phe^{11} form a second stacking pair.

Case V: L- to D-Conversions in LHRH Antagonists. Several years ago, it was observed
that the [des-His^2]-LHRH [3.126] and [D-Phe^2]-LHRH [3.127] both exhibited weak antagonist properties. Further, more potent antagonist analogs were then prepared which
had sequences [D-Phe^2,D-Trp^{3,6}]-LHRH [3.128] and [D-Phe^2,Pro^3,D-Phe^6 or D-Trp^6]-LHRH
[3.129]. This author examined these conformations by conformational energy calculations and found that all of the potent antagonists have low-energy structures whose
side-chain positions looked remarkably similar to LHRH [3.3]. The exception to this
stereo-equivalence was in the orientation of the pGlu^1 *cis* peptide. In order for this
group to point out (as was found for LHRH), the analogs [D-pGlu^1,D-Phe^2,D-Trp^{3,6}]-
LHRH or [D-pGlu^1,D-Phe^2,Pro^3,D-Trp^6 or D-Phe^6]-LHRH would be necessary [3.3]. Subsequent synthesis and physiological testing showed that the D-pGlu^1 analogs had
advantageous properties for antagonist activity over those of the equivalent L-pGlu^1
analogs [3.130]. Clearly, this is a case where knowledge of the conformation of the
LHRH molecule was valuable in designing a more potent analog. It should be noted
here that it was not necessary for the calculated structure to be absolutely correct for this kind of design decision to be made. One really only needed to know
the *consequences* of multiple D-residue substitution on the conformation in order to
come to the same conclusion. In all fairness, it should also be pointed out that
there was no way of knowing what modifications one should make to LHRH to design an
antagonist, even when a calculated conformation of the agonist was available. Further work along these lines is clearly needed before one will be able a priori to
jump from an agonist to an antagonist analog using only the knowledge of the conformation of the molecule. Some headway is being made along these lines for a series
of peptides designed to have specific functions [3.125].

b) *N-Methyl and Depsipeptide Results*

Case VI: Eledoisin and [(N-Me)Leu[7]]-LHRH. The eledoisin analog described here is a hexapeptide fragment with the sequence: H-Lys-Phe-Ile-Gly-Leu-Met-NH$_2$. It is known to have depressor activity nearly as high as that of natural eledoisin. In the work described here [3.131] each imino group in the amide bond was replaced systematic-ally with either a N-methyl or oxygen. The results of rabbit-blood depressor activ-ity on the modified hexapeptide showed the following:

1) N-methylation of the Ile[3] and Met[6] imino groups nearly eliminated depressor ac-tivity, while the oxygen replacement of the Met[6] imino also eliminated the activ-ity, and the Ile[3] depsipeptide analog retained ~10% activity.
2) N-methylation of the Gly[4] imino group doubled the activity, while the equivalent depsipeptide retained only ~30%.
3) Both N-methylation *and* oxygen substitution of the Leu[5] imino group enhanced ac-tivity to ~120%.
4) The depsipeptide of the Phe[2] imino group retained ~90% activity, while the equi-valent N-methyl analog retained only ~10% activity.

Using the results presented earlier for conformation preference of N-methyl and depsipeptide analogs, we can reasonably suggest the following conformation require-ments for the structure of the hexapeptide. These features may be seen in Fig.3.18 which was constructed solely from models and energy calculations using the above data:

1) The Phe[2] N-H bond is probably *cis* to the C^α-C^β bond of Phe[2] and *cis* to the Lys C^α-C^β bond, or *cis* to both.
2) The Ile[3] imide N-H bond is part of an α_R configuration at Phe[2], which would be bad for N-methylation, but only moderately poor for the depsipeptide.
3) The Gly[4] imide N-H bond is most probably part of an (+,-) conformation at Gly[4] and points away from the inner part of the molecule. The N-CH$_3$ would stabilize

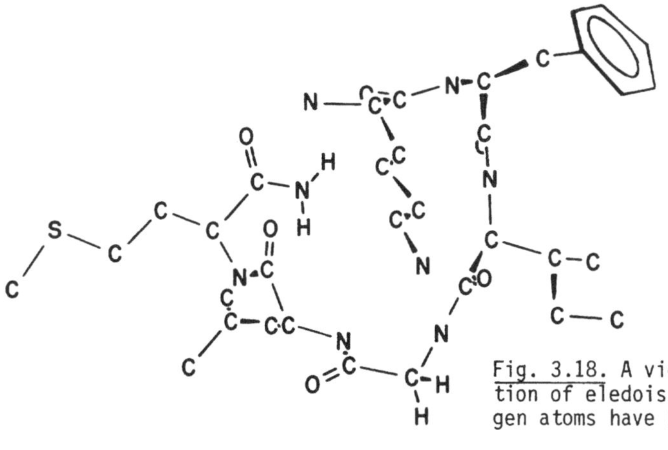

Fig. 3.18. A view of a conformer of a sec-tion of eledoisin. For clarity, some hydro-gen atoms have been omitted

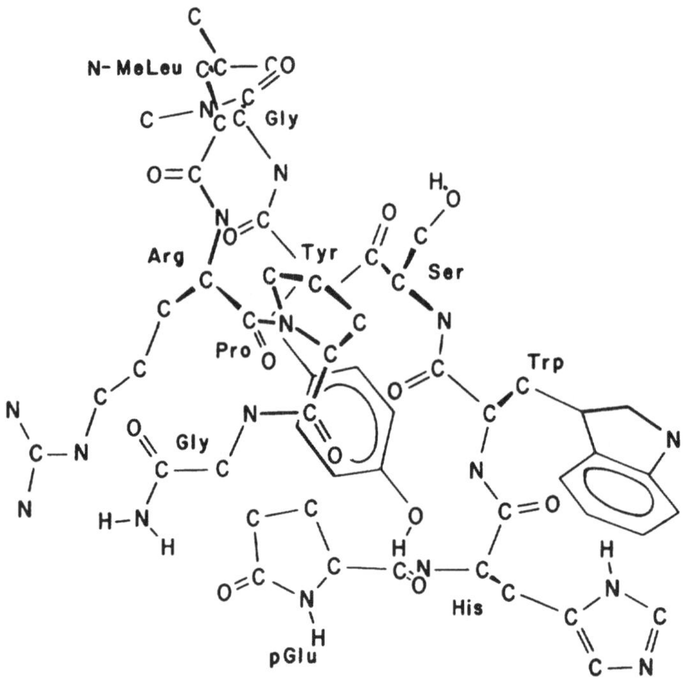

Fig. 3.19. A view of a conformer (CC) of [(N-Me)Leu^7]-LHRH. For clarity, some hydrogen atoms have been omitted

this part of a bend structure, while the depsipeptide would destabilize it somewhat.

4) The Leu^5 imino N-H is not involved in any hydrogen bonding, nor is its position modified by either substitution, thus, (-,+).

5) The Met^6 imino N-H is in the second position of a Gly (+,-), Leu (-,+) type II bend, but the α_R configuration around Met^6 would be destroyed by N-methylation.

The structure shown in Fig.3.18 reflects *one* possible conformation of this molecule and should only be considered as such. However, one can now predict, using this conformation, new analogs, which could further refine the model of the active conformation when combined with the above data. For example, this conformation suggests that D-residue substitution at the Gly^4 position could enhance activity.

The above analogs were all tested for proteolytic degradation with chymotrypsin and extended lifetimes were noted for the N-methylated analogs, while the depsipeptide degraded as rapidly as the native hexapeptide [3.131].

A particularly interesting example of the use of N-methylation to attempt to verify a proposed bend type can be found for LHRH [3.132]. In this case, the analog [(N-Me)Leu^7]-LHRH was designed to test for a hydrogen bond proposed to be between the N-H of Leu^7 and a carbonyl oxygen from Ser^4. The result was that the (N-Me)Leu^7 analog had ~100% activity, and further, [D-Ala^6,(N-Me)Leu^7]-LHRH had ~560% activity relative to native LHRH. Thus, the proposed [3.119] bend was disallowed and the bend structure, found by conformational energy calculations to be a modified type II, was to some extent justified [3.1,2]. Recent calculations [3.35] on the [(N-Me)Leu[7]]-LHRH analog confirmed that the predicted structure could remain intact when the N-methyl group was incorporated (Fig.3.19). A second bend conformation of the type α_L-α_L is allowed for [Gly[6],(N-Me)Leu[7]]-LHRH, but not for [D-Ala[6],(N-Me)Leu[7]]-LHRH [3.35].

c) L-, D-Combinations

Case VII: A Conformational Approach to the Design of Growth Hormone Agonist. Recently, it was observed that two synthetic enkephalin analogs, Tyr[1]-D-Trp[2]-Gly-Phe-Met-NH$_2$ and Tyr-D-Phe[2]-Gly-Phe-Met-NH$_2$, acted directly on the pituitary to release *only* GH [3.133]. The in vitro activity of these analogs was weak (1-100 µg/ml medium), but it was felt that by designing new analogs in part by empirical methods and more frequently by using calculated conformational data, we could achieve more active GH releasing factors [3.125]. Evidence that these analogs act via the receptor of natural GH-RH is presented elsewhere [3.133]. However, it is important to note that somatostatin, at low nanogram dosages, inhibits the release of GH activated by these analogs. This point will become important later in the discussion. Further, it was found that neither morphine, nor any of the potent [D-Ala[2]]-Met enkephalin-amide opiate-active analogs had in vitro GH releasing activity (C.Y. Bowers, unpublished result).

Conformational energy calculations [3.125] were carried out on the two GH releasing analogs and a schematic of the resulting structures is shown in Fig.3.20 for the D-Trp[2] analog. The structure shown in Fig.3.20 indicates that the L-Tyr[1]-D-Trp[2] rings

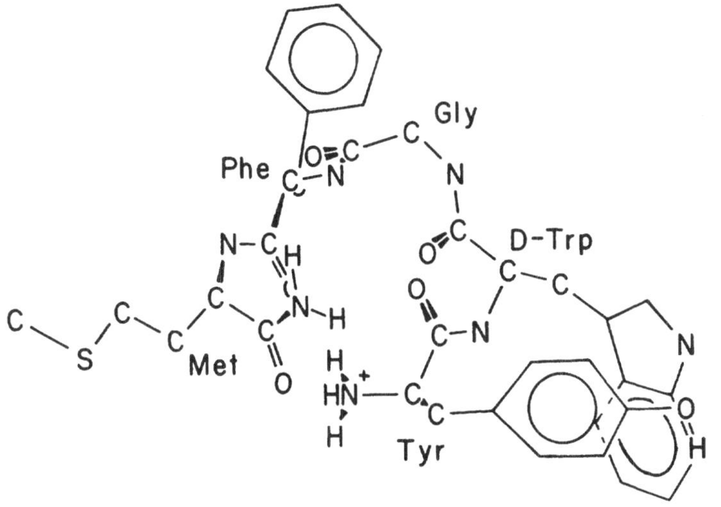

Fig. 3.20. A view of a conformer of [D-Trp[2]]-enkephalinamide

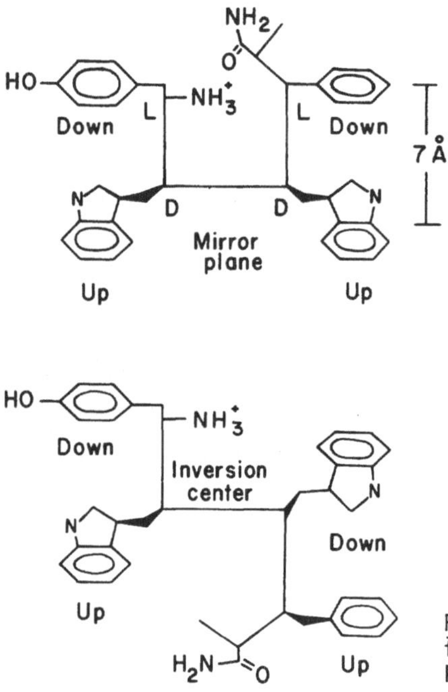

Fig. 3.21. A schematic diagram for possible configurations of the GH-RH analog, Tyr-D-Trp-D-Trp-Phe-Met-NH$_2$

can stack one above the other. It was further found by analog substitutions that GH releasing activity could be maintained if three of the four positions at the bend were ring-bearing residues and that highest activity occurred when L^1-ring, D^2-ring or D^3-ring, L^3-ring combinations were maintained [3.125]. For example, Tyr-D-Trp-Pro-Phe-Met-NH$_2$ was weakly active; as were Tyr-Ala-D-Phe-Lys-Met-NH$_2$ (3-position now D with 4-position L), Trp-D-Phe-D-Phe-Lys-Met-NH$_2$ (position 1 is now Trp^1 so Tyr^1 not a requirement for GH releasing activity, nor is Phe^4), and Tyr-D-Trp-Ser-Phe-D-Leu-NH$_2$ (position 3 allows L-configuration and position 5 allows D). By examining these analogs and others [3.125], it was determined that a L-D-D-L-X analog such as L-Tyr^1-D-Trp^2-D-Trp^3-L-Phe^4-Met^5-NH$_2$ should exhibit favorable GH releasing activity. The in vitro tests gave a potency for this analog at ~30 ng/ml medium, which is more potent than the original analogs and is approaching physiological activity. New analogs are currently being synthesized and tested which should increase the potency further. However, it is clear from the fact that 30 of 38 new analogs tested exhibited GH agonist activity that we now understand to some extent the structural criteria for interaction at the GH-RH receptor. The key structural effects are shown in Fig.3.21. At this time it is impossible to choose between the conformation with the approximate mirror plane and that with the approximate inversion center. However, the ring stacking distance is well-resolved as being the most important aspect for interaction at the GH-RH receptor. We remind the reader that in the discussion of SS, the D-Trp^8 analog was a very potent inhibitor to the release of GH. The author believes that the site of interaction for both agonists and inhibitor involves an intercalation of an aromatic ring from the receptor between the L-D-ring combinations described above. Clearly, one might extrapolate to suggest that the agonists bring *two* such groups together by *double* intercalation. Further work is being carried out to test this hypothesis for the mechanism of agonist/antagonist activity [3.125].

3.5 Conclusions

Evidence was presented here as to how one can use conformational information, synthesis, and physiological testing as an elegant and powerful procedure for rational drug design. Clearly, RUDINGER [3.134] had similar considerations in mind when he stated: "The problem of designing analogs of biologically active peptides is coextensive in its experimental aspects with the study of structure-activity relations. The difference is one of emphasis: whereas in a discussion of structure-activity relationships it is primarily the information gained in such studies which is of interest, a discussion of design will be concerned with the ways in which such information can best be obtained and in turn utilized for the guidance of further synthetic work". This author would go one step further and suggest the following procedure be followed for the most profitable polypeptide drug design. First, prepare analogs which contain *conformational* information within them (i.e., proline, N-methyl, α-methyl, D-substitution, etc.). Second, use the analog-activity data to derive a "hypothetical" structure. Third, probe the "active site" with subtle modifications made as nearly as possible either "isosteric" or "isofunctional". Clearly, this sequence is somewhat reversed from that normally carried out by most groups today. However, the importance of having a conformation of the molecule (whether from conformational energy calculations, experimental structural studies, appropriate analog studies, or a combination of all of these) cannot be overemphasized. A more efficient design program will ultimately allow the researcher to achieve his goals more quickly and economically.

The example cases described here have also shown the power of combining the three basic steps of conformation, synthesis, and activity testing. The place at which one starts in this triangle is not as important as how it is used. Hopefully, one can quickly spiral to the final drug with the desired properties. This author sees nothing but a bright future for those groups who can combine these three tools and efficiently use them in the design of new and more potent peptide drugs.

Acknowledgment. The author is indebted to the research personnel of the Bioproducts Division of Beckman Instruments, Inc. for much stimulating and helpful discussion during the writing of this chapter and to the Memphis State University Computing Center for valuable service.

References

3.1 F.A. Momany: J. Am. Chem. Soc. *98*, 2990 (1976)
3.2 F.A. Momany: J. Am. Chem. Soc. *98*, 2996 (1976)
3.3 F.A. Momany: J. Med. Chem. *21*, 63 (1978)
3.4 F.A. Momany: Biochem. Biophys. Res. Commun. *75*, 1098 (1977)
3.5 Y. Isogai, G. Nemethy, H.A. Scheraga: Proc. Natl. Acad. Sci. USA *74*, 414 (1977)
3.6 F.A. Momany, L.G. Drake, J.R. AuBuchon: Int. J. Quantum Chem. Symp. *5*, 381 (1978)
3.7 J.L. DeCoen, C. Humblet, M.H. Koch: FEBS Lett. *73*, 38 (1977)
3.8 M. Dygert, N. Go, H.A. Scheraga: Macromolecules *8*, 750 (1975)
3.9 G.H. Loew, S.K. Burt: Proc. Natl. Acad. Sci. USA *75*, 7 (1978)
3.10 J.S. Anderson, H.A. Scheraga: Macromolecules *11*, 812 (1978)
3.11 S. Fitzwater, Z.I. Hodes, H.A. Scheraga: Macromolecules *11*, 805 (1978)
3.12 B. Pullman: In *Quantum Mechanics of Molecular Conformations*, ed. by B. Pullman (John Wiley & Sons, New York 1976) p.295
3.13 S. Scheiner, C.W. Kern: J. Am. Chem. Soc. *100*, 7539 (1978)
3.14 P.N. Lewis, F.A. Momany, H.A. Scheraga: Isr. J. Chem. *11*, 121 (1973)
3.15 S.S. Zimmerman, M.S. Pottle, G. Nemethy, H.A. Scheraga: Macromolecules *10*, 1 (1977)
3.16 H.A. Scheraga: In *Peptides: Proceedings of the Fifth American Peptide Symposium*, ed. by M. Goodman, J. Meienhofer (John Wiley & Sons, New York 1977) p.246
3.17 H.A. Scheraga: In *Peptides, Polypeptides and Proteins*, ed. by E.R. Blout, F.A. Bovey, M. Goodman, N. Lotan (Wiley-Interscience, New York 1974) p.49
3.18 J. Snir, R.A. Nemenoff, H.A. Scheraga: J. Phys. Chem. *82*, 2497, 2527 (1978)
3.19 R.A. Nemenoff, J. Snir, H.A. Scheraga: J. Phys. Chem. *82*, 2504,2513, 2521 (1978)
3.20 F.A. Momany, L.N. Carruthers, R.F. McGuire, H.A. Scheraga: J. Phys. Chem. *78*, 1595 (1974);
     F.A. Momany, L.M. Carruthers, H.A. Scheraga: J. Phys. Chem. *78*, 1621 (1974)
3.21 F.A. Momany: J. Phys. Chem. *82*, 592 (1978)
3.22 F.A. Momany, R.F. McGuire, A.W. Burgess, H.A. Scheraga: J. Phys. Chem. *79*, 2361 (1975)
3.23 IUPAC-IUB Commission on Biochemical Nomenclature: J. Mol. Biol. *52*, 1 (1970)
3.24 A.W. Burgess, P.K. Ponnuswamy, H.A. Scheraga: Isr. J. Chem. *12*, 239 (1974)
3.25 M. Levitt: Biochemistry *17*, 4277 (1978)
3.26 M.B. Hossain, D. Vander Helm: J. Am. Chem. Soc. *100*, 5191 (1978)
3.27 I. Karle: J. Am. Chem. Soc. *100*, 1286 (1978)
3.28 I. Karle: In *Peptides*, ed. by M. Goodman, J. Meienhofer (John Wiley & Sons, New York 1977) p.274
3.29 I.L. Karle, J.W. Gibson, J. Karle: J. Am. Chem. Soc. *92*, 3755 (1970)
3.30 I.L. Karle: J. Am. Chem. Soc. *97*, 4379 (1975)
3.31 A.W. Burgess, S.J. Leach: Biopolymers *12*, 2599 (1973)
3.32 G.D. Smith, W.L. Duax, E.W. Czerwinski, N.E. Kendrick, G.R. Marshall, F.S. Mathews: In *Peptides*, ed. by M. Goodman, J. Meienhofer (John Wiley & Sons, New York 1977) p.277
3.33 J.L. Flippen, I.L. Karle: Biopolymers *15*, 1081 (1976)
3.34 A. Aubry, J. Protas, G. Boussard, M. Marrand, J. Neel: Biopolymers *17*, 1693 (1978)
3.35 P. Manavalan, F.A. Momany: Biopolymers *19*, 1943 (1980)
3.36 K. Inouye, K. Watanabe, K. Namba, H. Otsuka: Bull. Chem. Soc. Jpn. *43*, 3873 (1970)
3.37 E.C. Jorgensen, S.R. Rapaka, G.C. Windridge, T.C. Lee: J. Med. Chem. *14*, 899 (1971)
3.38 G.R. Marshall, N. Eilers, W. Vine: In *Progress on Peptide Research*, ed. by S. Laude (Gordon and Breach, New York 1972) p.15
3.39 G.R. Marshall, H.E. Bosshard, N.E. Kendrick, J. Turk, T.M. Balasubramanian, S.M.H. Cobb, M. Moore, L. Leduc, P. Needleman: In *Peptides 1976*, ed. by A. Loffet (Editions de l'Université de Bruxelles, Belgique 1976) p.361
3.40 J. Turk, P. Needleman, G.R. Marshall: J. Med. Chem. *18*, 1139 (1975)
3.41 R. Nagaraj, P. Balaram: FEBS Lett. *96*, 273 (1978)

3.42 A.W. Burgess, Y. Paterson, S.J. Leach: In *Peptides, Polypeptides and Proteins*, ed. by E.R. Blout, F.A. Bovey, M. Goodman, N. Lotan (John Wiley & Sons, New York 1974) p.79
3.43 J.C. Howard, F.A. Momany, R.H. Andreatta, H.A. Scheraga: Macromolecules *6*, 535 (1973)
3.44 A.E. Tonelli: Biopolymers *15*, 1615 (1976)
3.45 G.R. Marshall, F.A. Gorin: In *Peptides*, ed. by M. Goodman, J. Meienhofer (John Wiley & Sons, New York 1977) p.84
3.46 P. Groth: Acta Chem. Scand. *24*, 780 (1970)
3.47 P. Groth: Acta Chem. Scand. *A30*, 838, 840 (1976)
3.48 J. Konnert, I.L. Karle: J. Am. Chem. Soc. *91*, 4888 (1969)
3.49 Y. Iitaka, H. Nakamura, K. Takada, T. Takita: Acta Crystallogr. *B30*, 2817 (1974)
3.50 M. Goodman, F. Chen, C. Gilon, R. Ingwall, D. Nissen, M. Palumbo: In *Peptides, Polypeptides and Proteins*, ed. by E.R. Blout, F.A. Bovey, M. Goodman, N. Lotan (John Wiley & Sons, New York 1974)
3.51 M.J. Parry, A.B. Russell, M. Szelke: In *Chemistry and Biology of Peptides*, ed. by J. Meienhofer (Ann Arbor Science Publishing Inc., Ann Arbor, MI 1972) p.541
3.52 A.S. Dutta, B.J.A. Furr, M.B. Giles: In *Peptides*, ed. by M. Goodman, J. Meienhofer (John Wiley & Sons, New York 1977) p.189
3.53 A.S. Dutta, B.J.A. Furr, M.B. Giles, B. Valcaccia, A.L. Walpole: Biochem. Biophys. Res. Commun. *81*, 382 (1978)
3.54 A.S. Dutta, B.J.A. Furr, M.B. Giles, B. Valcaccia: J. Med. Chem. *21*, 1018 (1978)
3.55 G. Chipens, J. Ancan, G. Afanasyeva, J. Balodis, J. Indulen, V. Klusha, V. Kudryashova, E. Liepinsh, N. Makarova, N. Mishlyakova: In *Peptides 1976*, ed. by A. Loffet (Editions de l'Université de Bruxelles, Belgique 1976) p.353
3.56 A.S. Dutta, J.S. Morley: In *Peptides 1976*, ed. by A. Loffet (Editions de l'Université de Bruxelles, Belgique 1976) p.517
3.57 R. Chandrasekaran, A.V. Lakshminarayanan, U.V. Pandya, G.N. Ramachandran: Biochim. Biophys. Acta *303*, 14 (1973)
3.58 G.N. Ramachandran, R. Chandrasekaran: In *Progress in Peptide Research II*, ed. by S. Laude (Gordon and Breach, New York 1972) p.195
3.59 P.K. Ponnuswamy, V. Sasisekharan: Biopolymers *10*, 565 (1971)
3.60 C.M. Venkatachalam: Biopolymers *6*, 1425 (1968)
3.61 B. Maigret, B. Pullman: Theor. Chim. Acta *35*, 113 (1974)
3.62 P.N. Lewis, F.A. Momany, H.A. Scheraga: Proc. Natl. Acad. Sci. USA *68*, 2293 (1971)
3.63 P.N. Lewis, F.A. Momany, H.A. Scheraga: Biochim. Biophys. Acta *303*, 211 (1973)
3.64 J.C. Howard, A. Ali, H.A. Scheraga, F.A. Momany: Macromolecules *8*, 607 (1975)
3.65 S.S. Zimmerman, H.A. Scheraga: Proc. Natl. Acad. Sci. USA *74*, 4126 (1977)
3.66 K. Nishikawa, F.A. Momany, H.A. Scheraga: Macromolecules *7*, 797 (1974)
3.67 S.S. Zimmerman, L.L. Shipman, H.A. Scheraga: J. Phys. Chem. *81*, 614 (1977)
3.68 P.K. Ponnuswamy, P.K. Warme, H.A. Scheraga: Proc. Natl. Acad. Sci. USA *70*, 830 (1973)
3.69 E. Benedette: In *Peptides*, ed. by M. Goodman, J. Meienhofer (John Wiley & Sons, New York 1977) p.257
3.70 S.S. Zimmerman, H.A. Scheraga: Biopolymers *16*, 811 (1977)
3.71 G. Boussard, M. Marraud, J. Neel, B. Maigret, A. Aubry: Biopolymers *16*, 1033 (1977)
3.72 C. Lecomte, A. Aubry, J. Protas, G. Boussard, M. Marraud: Acta Crystallogr. *B30*, 1992, 1996, 2343, 2348 (1974)
3.73 M.M. Shemyakin, Y.A. Ovchinnikov, V.T. Ivanov: Angew. Chem. Int. Ed. Engl. *8*, 492 (1969)
3.74 C.F. Hayward, J.S. Morley: In *Peptides 1974*, ed. by Y. Wolman (John Wiley & Sons, New York 1975) p.287
3.75 F.A. Momany, J.R. Aubuchon: Biopolymers *17*, 2609 (1978)
3.76 M. Chorev, C.G. Willson, M. Goodman: In *Peptides*, ed. by M. Goodman, J. Meienhofer (John Wiley & Sons, New York 1977) p.572
3.77 C.G. Willson, M. Goodman, J. Rivier, W. Vale: In *Peptides*, ed. by M. Goodman, J. Meienhofer (John Wiley & Sons, New York 1977) p.579

3.78   A.C.M. Paiva, A.T. Ferreira, G. Goissis, T.P. Paiva: In *Peptides 1976*, ed.
       by A. Loffet (Editions de l'Université de Bruxelles, Belgique 1976) p.379
3.79   G. Goissis, V.L.A. Nouailhetas, A.C.M. Paiva: J. Med. Chem. *19*, 1287 (1976)
3.80   H.U. Immer, N.A. Abraham, V.R. Nelson, W.T. Robinson, K. Sestanj: In *Peptides
       1976*, ed. by A. Loffet (Editions de l'Université de Bruxelles, Belgique 1976)
       p.471
3.81   B.D. Hanlon: Beckman Bioproducts Division, private communication
3.82   J. Rudinger, K. Jost: Experientia *20*, 570 (1964)
3.83   J-L. Fauchere, N. Muthukumaraswamy, R. Schwyzer: In *Peptides 1976*, ed. by
       A. Loffet (Editions de l'Université de Bruxelles, Belgique 1976) p.647
3.84   V.M. Garsky, D.E. Clark, N.H. Grant: Biochem. Biophys. Res. Commun. *73*, 911
       (1976)
3.85   D. Sarantakis, J. Teichman, E.L. Lien, R.L. Fenichel: Biochem. Biophys. Res.
       Commun. *73*, 336 (1976)
3.86   D. Sarantakis, J. Reichman: In *Peptides*, ed. by M. Goodman, J. Meienhofer
       (John Wiley & Sons, New York 1977) p.186
3.87   D.F. Veber, R.G. Strachan, S.J. Bergstrand, F.W. Holly, C.F. Homnick,
       R. Hirschmann, M.L. Torchiana, R.J. Saperstein: J. Am. Chem. Soc. *98*, 2367
       (1976)
3.88   E.L. Lien, R.L. Fenichel, N.H. Grant, G.C. Boxill, J. Greenwood, J.P. Yardly:
       Biochem. Biophys. Res. Commun. *77*, 1317 (1977)
3.89   E. Gross, J.L. Morell: J. Am. Chem. Soc. *93*, 4634 (1971)
3.90   E. Gross, J.H. Brown: In *Peptides 1976*, ed. by A. Loffet (Editions de l'Uni-
       versité de Bruxelles, Belgique 1976) p.183
3.91   E. Gross, H.H. Kiltz, E. Nebelin: Hoppe-Seyler's Z. Physiol. Chem. *354*, 810
       (1973)
3.92   D.H. Rich, P.K. Bhatnagar: J. Am. Chem. Soc. *100*, 2212, 2218 (1978)
3.93   B.W. Bycroft: In *Chemistry and Biology of Peptides*, ed. by J. Meienhofer
       (Ann Arbor Science Publishing Inc., Ann Arbor MI 1972) p.665
3.94   J.C. Sheehan, D. Mania, S. Nakamura, J.A. Stock, K. Maeda: J. Am. Chem. Soc.
       90, 462 (1968)
3.95   E. Gross: In *Chemistry and Biology of Peptides*, ed. by J. Meienhofer (Ann
       Arbor Science Publishing, Inc., Ann Arbor, MI 1972) p.671
3.96   M.L. English, C.H. Stammer: Biochem. Biophys. Res. Commun. *85*, 780 (1978)
3.97   C.W. Smith, C.R. Botos, R. Walter: In *Peptides*, ed. by M. Goodman, J. Meien-
       hofer (John Wiley & Sons, New York 1977) p.161
3.98   C.W. Smith, R. Walter: Science *199*, 297 (1978)
3.99   G.H. Fisher, D.I. Marlborough, J.W. Ryan, A.M. Felix: Arch. Biochem. Biophys.
       189, 81 (1978)
3.100 R. Boni, R. DiBlasi, A. Farina, A.S. Verdini: Biopolymers *15*, 1233 (1976)
3.101 H.M. Berman, E.L. McGandy, J.W. Burgner, R.L. VanEtten: J. Am. Chem. Soc. *91*,
       6177 (1969)
3.102 T. Takeuchi, D.J. Prockop: Biochim. Biophys. Acta *175*, 142 (1969)
3.103 T. Takeuchi, J. Rosenbloom, D.J. Prockop: Biochim. Biophys. Acta *175*, 156
       (1969)
3.104 C.G. Willson, C. Gilon, B. Donzel, M. Goodman: Biopolymers *15*, 2317 (1976)
3.105 E. Benedetti, A. Christensen, C. Gilon, W. Fuller, M. Goodman: Biopolymers
       15, 2523 (1976)
3.106 D. Yamashiro, L. Fu Tseng, C.H. Li: Biochem. Biophys. Res. Commun. *78*, 1124
       (1977)
3.107a F.A. Momany, P. Manavalan: in preparation
3.107b E. Benedetti, B. DiBlasio, V. Pavone, C. Pedone, A. Felix, M. Goodman: Bio-
       polymers *20*, 283 (1981)
3.108 C. Meyers, A. Arimura, A. Gordin, R. Fernandez-Durango, D.H. Coy, A.V. Schally,
       J. Drouin, L. Ferland, M. Beanlien, F. Labrine: Biochem. Biophys. Res. Commun.
       74, 630 (1977)
3.109 Y. Hirotsu, D.H. Coy, A.V. Schally: Biochem. Biophys. Res. Commun. *59*, 277
       (1974)
3.110 W. Arnold, G. Flouret, R. Morgan, R. Rippel, W. White: J. Med. Chem. *17*,
       314 (1974)
3.111 D.H. Coy, A.J. Kastin, A.V. Schally, O. Morin, N.G. Caron, F. Labrie, J.M.
       Walker, R. Fertel, G.G. Berntson, C.A. Sandman: Biochem. Biophys. Res. Commun.
       73, 632 (1976)

3.112 C.B. Pert, A. Pert, J.K. Chang, B.T.W. Fong: Science *194*, 330 (1976)
3.113 C.R. Beddell, R.B. Clark, G.W. Hardy, L.A. Lowe, F.B. Ubatuba, J.R. Vane,
      F.R.S. Wilkinson, K.J. Chang, P. Cuatrecasas, R.J. Miller: Proc. Roy. Soc.
      London Ser. B *198*, 249 (1977)
3.114 J.M. Hambrook, B.A. Morgan, M.J. Ranee, C.F.C. Smith: Nature (London) *262*,
      782 (1976)
3.115 R.J. Miller, K. Chang, P. Cuatrecasas, S. Wilkinson: Biochem. Biophys. Res.
      Commun. *74*, 1311 (1977)
3.116 N. Ling, S. Minick, L. Lazarus, J. Rivier, R. Guillemin: In *Peptides*, ed. by
      M. Goodman, J. Meienhofer (John Wiley & Sons, New York 1977) p.96
3.117 J. Hughes, T.W. Smith, H.W. Kosterlitz, L.A. Fothergill, B.A. Morgan, H.R.
      Morris: Nature (London) *258*, 577 (1975)
3.118 P. Manavalan, F.A. Momany: "A Comparison of Low Energy Structures of Enkephalin
      Analogs", in *Peptides: Structure and Biological Function*, ed. by E. Gross and
      J. Meienhofer (Pierce Chemical Co., Rockford, IL 1979) p.893
3.119 M.W. Monahan, M. Amoss, H. Anderson, W. Vale: Biochemistry *12*, 4616 (1973)
3.120 M. Fujino, I. Yamazaki, S. Kobayashi, T. Fukuda, S. Shinagana, R. Nakayama,
      W.F. White, R.H. Rippel: Biochem. Biophys. Res. Commun. *57*, 1248 (1974)
3.121 D. Coy, F. Labrie, M. Savary, E. Coy, A.V. Schally: Biochem. Biophys. Res.
      Commun. *67*, 576 (1975)
3.122 W. Vale, C. Rivier, M. Brown, J. Leppaluoto, N. Ling, M. Monahan, J. Rivier:
      Clin. Endocrinol. *5*, 261 (1976)
3.123 J. Rivier, M. Brown, C. Rivier, N. Ling, W. Vale: In *Peptides 1976*, ed. by
      A. Loffet (Editions de l'Université de Bruxelles, Belgique 1976) p.427
3.124 J. Rivier, M. Brown, W. Vale: Biochem. Biophys. Res. Commun. *65*, 746 (1975)
3.125 F.A. Momany, C.Y. Bowers, G.A. Reynolds, D. Chang, A. Hong, K. Newlander:
      "Design, Synthesis, and Biological Activity of Peptides which Release Growth
      Hormone *in vitro*", in Endocrinology *108*, 31 (1981)
3.126 W. Vale, G. Grant, J. Rivier, M. Monahan, M. Amoss, R. Blackwell, R. Burgus,
      R. Guillemin: Science *176*, 933 (1972)
3.127 R. Rees, T. Foell, S. Chai, N. Grant: J. Med. Chem. *17*, 1016 (1974)
3.128 D. Coy, J. Vilchez-Martinez, A.V. Schally: In *Peptides 1976*, ed. by A. Loffet
      (Editions de l'Université de Bruxelles, Belgique 1976) p.463
3.129 J. Humphries, Y.-P. Wan, K. Folkers, C.Y. Bowers: Biochem. Biophys. Res.
      Commun. *72*, 939 (1976)
3.130 J.E. Rivier, W.W. Vale: Life Sci. *23*, 869 (1978)
3.131 M. Miyoshi, H. Sugano: In *Peptides 1974*, ed. by Y. Wolman (John Wiley & Sons,
      New York 1975) p.355
3.132 N. Ling, W. Vale: Biochem. Biophys. Res. Commun. *63*, 801 (1975)
3.133 C.Y. Bowers, G.A. Reynolds, D. Chang, A. Hong, K. Chang, F.A. Momany: Endo-
      crinology *108*, 1071 (1981)
3.134 J. Rudinger: In *Drug Design*, Vol.2, ed. by E.J. Ariens (Academic Press, New
      York 1971) p.319

4. Cohesion and Ionicity in Organic Semiconductors and Metals

R. M. Metzger

With 7 Figures

The cohesive energy calculations for organic ionic crystals use quantum-chemical atom-in-molecule partial charges, dipole moments and polarizabilities, and classical lattice sums. These calculations can be connected with Löwdin's formal, fully quantum-mechanical theory and with the relevant Born-Haber cycles.

The Madelung (or charge-charge) energy dominates the cohesive energy for organic ionic insulators and semiconductors. For organic partially ionic quasi-one-dimensional metals like TTF TCNQ, both the Madelung energy and the uniform partially charged lattice model fail to account for the existence of the crystal; polarization, dispersion, and charge-dipole energies become very important, and they can help stabilize the partially ionic lattice either with the over-idealized Wigner lattice or with a nonlinear expression of the cost of partial ionization.

4.1 Introductory Comments

The detailed theoretical study of the cohesive energy of *neutral* hydrocarbons by the atom-atom interaction model has been pioneered by KITAIGORODSKY and coworkers [4.1-12] and WILLIAMS and coworkers [4.13-22]. As is reviewed in Chap.2, one starts from a very restricted but representative data base of experimental thermochemical data (enthalpies of formation and of sublimation) and crystallographic data (room-temperature, and if available, low-temperature crystal structure determinations). From these data for "model compounds" a self-consistent set of semiempirical "nonbonded atom parameters" can be constructed for C, H, and even N, O, and S atoms in neutral aliphatic and aromatic crystals.

Initially, only the dispersion energy E_d and the repulsion energy E_r were considered in a "lumped-parameter" description of organic cohesion. Later a relatively small Coulomb, or Madelung, contribution was added to help predict the experimental crystal structure, and for specific bonding situations a Morse potential has been considered (Chap.2).

For $C_x H_y$ crystals, these calculations are highly successful in predicting crystal structure ("packing") and have sparked a parallel effort to understand, again semi-

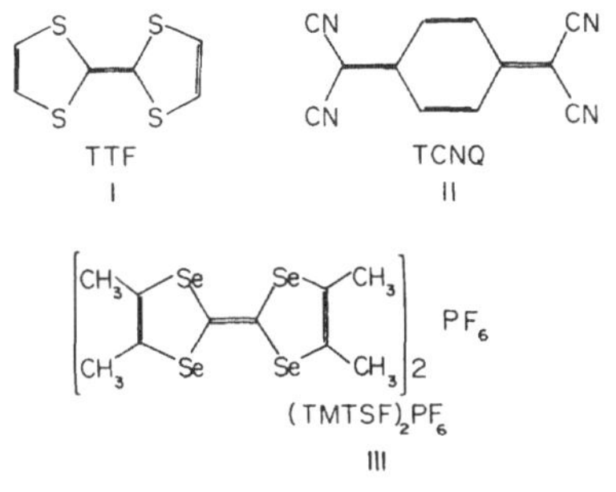

Fig. 4.1. The complex, TTF TCNQ, is a quasi-one-dimensional metal, while $(TMTSF)_2PF_6$ superconducts at low temperature and high pressure

empirically, the most stable biological conformations for amino acids, peptides and biopolymers in general. Here again, as reviewed in Chap.3, a large measure of success and acceptance has been achieved. Even for heteroatom crystals the packing parameters for O, N, and S are becoming reliable. In general, it seems that the goal of predicting the crystal conformation of any neutral organic crystal with limited computing effort is within reach.

For organic ionic and partially ionic crystals the situation is not as ideal. Naturally, the domain of existence of these crystals is much more restricted (for reviews, cf. [4.23-33]), but a thermochemical data base is almost non-existent for them [4.34,35]. Whereas over twenty years of research effort have been dedicated to the study of neutral organic compounds [4.1-22], the study of Madelung and other cohesive energy contributions for organic ionic crystals is only eight years old [4.36-64]. Nevertheless, the existence of quasi-one-dimensional metals (QODM), e.g., TTF TCNQ (tetrathiafulvalene(I), 7,7,8,8-tetracyanoquinodimethan(II)) [4.65,66] (see Fig.4.1) as a subset of the partially ionic organic crystals, and the discovery of superconductivity in $(TMTSF)_2PF_6(III)$ [4.67] have spurred a considerable effort to understand the cohesion and, hopefully, to predict the fractional ionicity in these compounds.

The present chapter follows fairly closely an earlier summary [4.41] of the results up to mid-1977, but the advances made since then seem to warrant a new review.

Section 4.2 provides a general introduction to the problem of crystalline cohesion. Section 4.3 reviews the various forms of the Born-Haber cycle applicable to organic ionic crystals and discusses prevailing criteria for ionicity. The results of most calculations, including for completeness those already reviewed in [4.41], are given in Sect.4.4.

4.2 Crystal Cohesive Energies: General Theory

4.2.1 Löwdin's Theory

All cohesive energy calculations reviewed here are based on atomic parameters (fractional charge, local dipole moment, polarizability, etc.) localized at atomic positions in the crystal structure, i.e., they are inspired by the atom-in-molecule approximation introduced by KITAIGORODSKY [4.1,2]. Nonetheless, it is important that these calculations be connected with the more elegant, fully quantum-mechanical formalism for crystalline cohesion, pioneered mainly by LÖWDIN [4.68]. This connection is presented here in the hope that some useful insights might be obtained.

According to LÖWDIN [4.68] the theoretical cohesive energy U_{coh} may be defined as the difference between the total energy of the crystal E_t and the energy E_f of its N individual "free components" (atoms, molecules, or ions) in the gas phase:

$$U = E_t - E_f \ . \tag{4.1}$$

As is well known, within the confines of the Born-Oppenheimer approximation, the zeroth-order Hartree-Fock estimate of E_f (call it $E_f^{(0)}$) can be decomposed for each component into the sum

$$E_f^{(0)} = N(E_{f,NN} + E_{f,DC} + E_{f,EC} + E_{f,NA} + E_{f,KE}) \ , \tag{4.2}$$

where $E_{f,NN}$, $E_{f,DC}$, $E_{f,EC}$, $E_{f,NA}$, and $E_{f,KE}$ are, respectively, the classical nucleus-nucleus repulsion, the direct Coulomb electron-electron repulsion, the exchange Coulomb electron-electron repulsion, the electron-nucleus attraction, and the electronic kinetic energy. The London dispersion (induced dipole-induced dipole) contributions to E_f are usually obtained from second-order perturbation theory calculations for the free component and are thus considered as part of a "second-order" contribution to E_f (call it $E_f^{(2)}$):

$$E_f = E_f^{(0)} + E_f^{(2)} + \dots \ . \tag{4.3}$$

The exact quantum-mechanical calculation of E_t

$$E_t = E_t^{(0)} + E_t^{(2)} + \dots \ , \tag{4.4}$$

or even of its zeroth-order estimate $E_t^{(0)}$ is highly nontrivial. $E_t^{(0)}$ has been calculated for alkali halides [4.68] and for metallic hydrogen [4.69]. Of course, approximate parametrized treatments abound (APW, OPW, Wigner-Seitz, Xα, etc.) and have reached relatively high levels of sophistication, at least for simple metals [4.70]. For molecules of the size of TTF or TCNQ, the semiempirical NDO (neglect of differential overlap) [4.71] theories are quite practical for calculations of $E_f^{(0)}$ (and perhaps even $E_f^{(2)}$ [4.72]). Of these NDO theories, CNDO/2 [4.73] has been used in a brute-force direct-space extension [4.74] to provide estimates of $E_t^{(0)}$ without using Bloch crystal orbitals, obtaining $E_t^{(0)}$ expressions that can be recast into a tight-binding formalism [4.74]. However, this method has not been used very widely.

The usual formal approach to $E_t^{(0)}$, at least for organic crystals, is to construct Bloch-symmetrized crystal orbitals from the undistorted single-molecule molecular orbitals [4.75] and to evaluate $E_t^{(0)}$ from them. In principle, one should then solve the crystal SCF matrix in the tight-binding formalism to minimize $E_t^{(0)}$. It can be shown formally [4.75] that a decomposition identical to that of (4.2) can be achieved for $E_t^{(0)}$

$$E_t^{(0)} = E_{t,NN} + E_{t,DC} + E_{t,EC} + E_{t,NA} + E_{t,KE} \quad , \tag{4.5}$$

provided that either the intermolecular overlap is accounted for by brute-force Löwdin crystal orbital orthonormalization [4.76], or else the overlap between crystal unit cells is set to zero for the crystal orbitals that belong to the same Bloch wave vector, but come from different zeroth cell molecular orbitals [4.75]. Since each term in (4.5) contains lattice sums, the calculation of $E_t^{(0)}$ is not easy. Thus, until practical algorithms based on (4.5) become routinely available, the semiclassical, semiempirical methods discussed below will retain their usefulness.

A correlation can be established between the several semiclassical lattice energies introduced below and their counterparts in Löwdin's formulation of the cohesive energy. LÖWDIN has shown [4.68] that if the n_g wave functions $\{\Psi_\lambda(r), \lambda=1,2,\dots,n_g\}$ are used to define the "free" electron density ρ_g for the "free" (vapor phase) constituent g (i.e., the single molecule g with its vapor-phase equilibrium geometry), and if this density becomes $\bar{\rho}_g$ in the crystal (due to Bloch-symmetrized crystal orbitals $\Psi_{\lambda\omega t}(r)$ [4.75]), then one may define the electron density shift $\Delta\rho_g$ as

$$\Delta\rho_g \equiv \bar{\rho}_g - \rho_g \quad . \tag{4.6}$$

Then the cohesive energy (to zeroth order) becomes

$$U^{(0)} \equiv E_t^{(0)} - E_f^{(0)} \tag{4.7a}$$

$$= E_{elstat} + E_{exch} + E_S \quad , \tag{4.7b}$$

where E_{elstat} is the quantum-mechanical analog of the classical electrostatic energy for the "undistorted free constituents", E_{exch} is the intermolecular exchange Coulomb energy for the "undistorted free constituents", and E_S is the overlap energy. This overlap energy consists of all the terms due to the solid-state density deformations $\Delta\rho_g$ which in turn reflect the effects due to the Löwdin crystal overlap integrals S [4.76].

Let the crystal consist of N molecules (N is usually Avogadro's number), and let each unit cell consist of Z molecules. Denote the crystal lattice translation vector by

$$\underline{r}_L = n_1\underline{a} + n_2\underline{b} + n_3\underline{c} \quad , \quad (L=0,1,\ldots,N/Z) \tag{4.8}$$

where $\underline{a}$, $\underline{b}$, $\underline{c}$ are the crystallographic unit cell axes, and n_1, n_2, n_3 are integers. Then a lattice sum is represented by $\sum_{L=0}^{N/Z}$, and E_{elstat} can be decomposed as

$$E_{elstat} = E_{NN} + E_{NA} + E_{DC} \quad , \tag{4.9}$$

where NN, NA, DC have the same meaning as in (4.2). Further

$$E_{NN} \equiv E_{t,NN} - NE_{f,NN} = \frac{N}{Z} \sum_{g \leq h}{}' \sum_{L=0}^{N/Z} \frac{Z_g Z_h}{|\underline{r}_g - \underline{r}_h - \underline{r}_L|} \quad , \tag{4.10}$$

$$E_{NA} = -\frac{N}{Z} \sum_{g \neq h} \sum_{L=0}^{N/Z} Z_g \int \frac{\rho_h(1,1)dv_1}{|\underline{r}_g - \underline{r}_h - \underline{r}_L|} \quad , \tag{4.11}$$

$$E_{DC} = \frac{N}{Z} \sum_{g < h} \sum_{L=0}^{N/Z} \iint \frac{\rho_g(1,1)\rho_{h+L}(2,2)dv_1 dv_2}{|\underline{r}_1 - \underline{r}_2|} \quad , \tag{4.12}$$

where $\rho_h(1,1) = \psi_h^*(1)\psi_h(1)$ is the electron density function, Z_g is the nuclear charge of constituent g, and $\underline{r}_g$ is the position vector of constituent g. The prime in the summation over h in (4.10) indicates that the singular case $g = h = L = 0$ is not included in the summation.

The next term in (4.7b) is the exchange energy,

$$E_{exch} = -\frac{N}{Z} \sum_{g < h} \sum_{L=0}^{N/Z} \iint \frac{\rho_g(1,2)\rho_{h+L}(2,1)dv_1 dv_2}{|\underline{r}_1 - \underline{r}_2|} \quad , \tag{4.13}$$

for which there is no classical counterpart.

The last term in (4.7b), the overlap energy E_S, can be written [4.68] as

$$E_S = -\frac{N}{Z} \sum_{g \neq h} \sum_{L=0}^{N/Z} Z_g \int \frac{\Delta\rho_h(1,1)dv_1}{|\underline{r}_g - \underline{r}_h - \underline{r}_L|}$$

$$+ \frac{N}{Z} \sum_{g \neq h} \sum_{L=0}^{N/Z} \iint \frac{[\Delta\rho_g(1,1)\Delta\rho_{h+L}(2,2) - \Delta\rho_g(1,2)\Delta\rho_{h+L}(2,1)]dv_1 dv_2}{|\underline{r}_1 - \underline{r}_2|}$$

$$+ \frac{N}{2Z} \sum_g \iint \frac{\Delta\rho_g(1,1)\Delta\rho_g(2,2) - \Delta\rho_g(1,2)\Delta\rho_g(2,1)}{|\underline{r}_1 - \underline{r}_2|} dv_1 dv_2$$

$$+ \frac{N}{Z} \sum_g \int \left[\frac{\nabla_1^2}{2m\hbar^2} - \frac{Z_g}{|\underline{r}_g - \underline{r}_1|}\right] \Delta\rho_g(1,1') \bigg|_{x'=x_1} dv_1 \quad . \tag{4.14}$$

The last term of the right-hand side of (4.14) contains the kinetic energy difference $E_{t,KE} - E_{f,KE}$. It has been shown by LANDSHOFF [4.77] that if the free constituents of the crystal are closed-shell ions and the crystal is an insulator, then

$$E_{t,KE} - E_{f,KE} = 0 \quad . \tag{4.15}$$

Also, (4.15) holds if zero intercell overlap is used [4.75], in which case metallic band properties will vanish. However, for broad-band metals much of the crystal cohesion comes from the fact that $E_{t,KE} - E_{f,KE}$ is far from zero. For narrow-band metals this kinetic energy contribution is small [4.49]. However, even for insulators such as the alkali halides, the full expression (4.14) for the overlap energy E_S is far from zero [4.68]. In fact, the physical justification for the use of "ad hoc" repulsion energies E_r (V_r in Chap.2) [4.1-22] is that they are a classical approximation to E_S.

There is, in addition to E_{elstat}, E_{exch}, and E_S, a zero-point kinetic energy contribution, E_{ZPKE}, which is required by the Pauli exclusion principle but is rather small [4.68].

4.2.2 Madelung Energy

The classical cohesive energies [4.35-64] reviewed here apply the atom-in-molecule model [4.1,2], but rely on quantum-chemical calculations for the isolated molecule for the necessary fundamental parameters (charge, dipole moment, polarizability). The connection between these cohesive energy calculations and Löwdin's elegant formalism [4.68] will now be made.

Of the four contributions to $U^{(0)}$,

$$U^{(0)} = E_{elstat} + E_{exch} + E_S + E_{ZPKE} \quad , \tag{4.16}$$

E_{ZPKE} may be safely neglected since it is small, and E_{exch} must unfortunately be ignored (even though it may be very important) because it cannot be obtained from a classical lattice sum.

One may expand E_{elstat} in a classical multiple expansion [4.78]:

$$E_{elstat} = E_{cc} + E_{cd} + E_{cq} + \cdots + E_{dd} + E_{dq} + \cdots \tag{4.17}$$

(c=charge, d=permanent electric dipole, q= permanent quadrupole). To provide a classical treatment of Löwdin's E_{elstat} one may define an electron density function $\rho(r)$

$$\rho(\underline{r}) = \sum_{L=0}^{N/Z} \sum_{g} [Z_g \delta(\underline{r}-\underline{r}_g-\underline{r}_L) - \rho_g(\underline{r}-\underline{r}_g-\underline{r}_L)] \quad , \tag{4.18}$$

where the sum over g extends only over the molecules in the zeroth unit cell, and $\delta(\underset{\sim}{r})$ is the Dirac delta function. Then the exact classical expression is

$$E_{elstat} = \frac{1}{2} \iint \frac{\rho(\underset{\sim}{r})\rho(\underset{\sim}{r}+\underset{\sim}{R})}{|\underset{\sim}{R}|} \, dv(\underset{\sim}{r})dv(\underset{\sim}{R}) \quad . \tag{4.19}$$

If one approximates $\rho_g(\underset{\sim}{r})$ by a spin density, or better, by the Mulliken gross atomic population $-q_g + Z_g$, then the continuous electron density $\rho(\underset{\sim}{r})$ of (4.19) simplifies to

$$\rho(\underset{\sim}{r}) = \sum_{L=0}^{N/Z} \sum_{i=1}^{M} q_i \delta(\underset{\sim}{r}-\underset{\sim}{r}_i-\underset{\sim}{r}_L) \quad , \tag{4.20}$$

and one deals now with net fractional point charges q_i; the sum over i extends to the M atoms in the zeroth unit cell. Then E_{elstat} reduces to the first term of the right-hand side of (4.17), the charge-charge or Madelung energy E_M [4.36],

$$E_M = E_{cc} = \frac{N}{Z} \sum_{L=0}^{N/Z} \sum_{i \geq j=1}^{M} \sum_{}^{M'} \frac{q_i q_j}{|\underset{\sim}{r}_i-\underset{\sim}{r}_j-\underset{\sim}{r}_L|} \quad , \tag{4.21}$$

where $\underset{\sim}{r}_i$ is the position of atom i in the zeroth cell. This Madelung energy can also be expressed [4.39,40,47,53,54] in terms of Madelung site potentials,

$$E_M = N \sum_{i=1}^{M} q_i \phi_i^M(\underset{\sim}{r}_i) \quad , \tag{4.22}$$

where the site potential $\phi_i^M(\underset{\sim}{r})$ is given by

$$\phi_i^M(\underset{\sim}{r}_i) \equiv \frac{1}{2Z} \sum_{L=0}^{N/Z} \sum_{j=1}^{M'} \frac{q_j}{|\underset{\sim}{r}_i-\underset{\sim}{r}_j-\underset{\sim}{r}_L|} \quad . \tag{4.23}$$

This site potential may be differentiated to yield the Madelung electric field $F_i^M(\underset{\sim}{r})$:

$$F_i^M(\underset{\sim}{r}_i) = -\nabla \phi_i^M(\underset{\sim}{r}_i) = \frac{1}{2Z} \sum_{L=0}^{N/Z} \sum_{j=1}^{M'} \frac{q_j(\underset{\sim}{r}_i-\underset{\sim}{r}_j-\underset{\sim}{r}_L)}{|\underset{\sim}{r}_i-\underset{\sim}{r}_j-\underset{\sim}{r}_L|^3} \quad . \tag{4.24}$$

4.2.3 Charge-Dipole and Dipolar Energies

In some cases, as is explained below, the Madelung energy is not sufficient to domi-nate the cohesive energy, and one may wish to look beyond (4.20) to find a better approximation for $\rho(\underset{\sim}{r})$. (The other approach is to include "second-order" contribu-tions due to atom polarizabilities: this is discussed in Sect.4.2.4.)

An observable related to the molecular electron density $\rho_g(\underset{\sim}{r})$ is the molecular electric dipole moment $\underset{\sim}{\mu}$. Its quantum-mechanical calculation, based on the molecular density matrix, yields three terms [4.79]:

$$\underset{\sim}{\mu} = \underset{\sim}{\mu}^{chg} + \underset{\sim}{\mu}^{hyb} + \underset{\sim}{\mu}^{asym} \quad . \tag{4.25}$$

The asymmetry dipole vanishes for the NDO quantum-chemical calculations used for cohesive energy work [4.79,80] because of the NDO parametrization. The net charge dipole μ^{chg},

$$\mu^{chg} = 4.80325 \sum_{i=1}^{A} q_i r_i \quad , \tag{4.26}$$

uses the same net atomic charges q_i that also enter into the Madelung energy expression (4.21). The sum in (4.26) is over the A atoms in the molecule. The hybridization dipole μ^{hyb} corrects for the distortion from spherical symmetry of the atomic wave functions, expressed as the Pauling hybridization, e.g., of the s and p wave functions. This hybridization dipole uses [4.71,73,79,80] one-center off-diagonal elements of the molecular density matrix which are not used in (4.21,26).

Thus one may bring the atom-in-molecule contributions μ_i^{hyb} [4.80] together with the fractional point charges q_i to yield a better $\rho(r)$. Following STAKGOLD [4.81] one may define [4.40]

$$\rho(r) = \sum_{L=0}^{N/Z} \sum_{i=1}^{M} q_i \delta(r - r_i - r_L) + \mu_i^{hyb} \cdot \nabla \delta(r - r_i - r_L) \quad . \tag{4.27}$$

By substituting (4.27) in (4.19), E_M is obtained as in (4.21), and also

$$E_{cd} \equiv -N \sum_{i=1}^{M} \mu_i^{hyb} \cdot F_i^M(r_i) \quad , \tag{4.28}$$

$$E_\mu \equiv E_{dd} \equiv \frac{N}{Z} \sum_{L=0}^{N/Z} \sum_{i \geq j=1}^{M} \sum^{M'} \left[\frac{\mu_i^{hyb} \cdot \mu_j^{hyb}}{|r_i - r_j - r_L|^3} - \frac{3 \mu_i^{hyb} \cdot (r_i - r_j - r_L) \mu_j^{hyb} \cdot (r_i - r_j - r_L)}{|r_i - r_j - r_L|^5} \right] \quad , \tag{4.29}$$

where E_μ is the dipolar energy [4.2] or permanent hybrid dipole-permanent hybrid dipole interaction energy, and E_{cd} is a "cross" term that includes interactions between fractional atomic charges on any molecule with local moments on other molecules in the crystal.

With (4.27), therefore, all terms involving charges and permanent dipoles in (4.17) have been found for E_{elstat}.

Naturally, all intramolecular contributions must be corrected for in E_M, ϕ_i^M, F_i^M, and E_μ (4.21,23,24,29). The presumably important contributions to E_{elstat} having been identified, the remaining term of interest in (4.16) is E_S. The discussion that follows (4.14) justifies the expression

$$E_S = E_r + E_k \quad , \tag{4.30}$$

where E_r is the repulsion energy of the form

$$E_r = \frac{N}{Z} \sum_{L=0}^{N/Z} \sum_{i \geq j=1}^{M} \sum^{M} \frac{R_i R_j}{|r_i - r_j - r_L|^n} \quad , \tag{4.31}$$

with $n > 9$ and R_i and R_j being "ad hoc" adjustable Born-Landé parameters, or

$$E_r = \frac{N}{Z} \sum_{L=0}^{N/Z} \sum_{i \geq j=1}^{M} \sum^{M} B_i B_j \exp(-C_{ij}|\underline{r}_i - \underline{r}_j - \underline{r}_L|) \quad , \tag{4.32}$$

where again B_i, B_j, C_{ij} are adjustable parameters (Chap.2) which could conceivably be different [4.38] for neutral and ionic crystals. The energy E_r is crucial in preventing the collapse of the crystal lattice to a single point under the influence of the Coulomb force. The next term in (4.30), E_k, is the kinetic energy term, which includes all metallic band effects [4.49]. It should also be noted that since most cohesive energy calculations start from the published crystal structure and compute atom charges q_i and local moments μ_i^{hyb} from the molecular geometry in the crystal and not in vacuo, therefore some part of Löwdin's E_S contributions (4.14) is already included in the calculation of E_M, E_{cd}, and E_μ.

At this point it seems appropriate to reply to an objection [4.82] and reiterate the justification of all Born-Mayer-type crystal cohesive energy calculations in light of the requirements of the virial theorem. Both for the free constituents at infinite mutual separation and for the crystal at equilibrium the virial theorem must be satisfied. In a Coulomb field the Born-Mayer cohesive energy U is given by

$$U = T + V \quad , \tag{4.33}$$

where the crystal kinetic energy T [not to be confused with (4.15)] is

$$T = -U \quad , \tag{4.34}$$

and because of the r^{-1} Coulomb potential, the crystal potential energy V is

$$V = 2U \quad . \tag{4.35}$$

FROMAN and LÖWDIN [4.83] have pointed out that the virial theorem is satisfied, provided that (4.15) is not used, and that, for ionic crystals, an additional potential energy V_{ext}, separate from the usual Madelung energy E_M and due to the "extension" of the ions, is used:

$$V = V_M + V_{ext} \quad . \tag{4.36}$$

Furthermore, if an "ad hoc" repulsion energy E_r [(4.31) or (4.32)] is used as an approximation to V_{ext}, then again, the virial theorem is satisfied [4.84].

4.2.4 Polarization and Dispersion Energies

The constituents of $U^{(0)}$ (4.7) having been discussed, the "second-order" energy

$$U^{(2)} = E_t^{(2)} - E_f^{(2)} \quad , \tag{4.37}$$

must next be considered. $U^{(2)}$ includes all interactions that in molecular calculations would be obtained by second-order perturbation theory. Thus, $U^{(2)}$ consists of the polarization energy E_{pol}, the London dispersion energy E_d, and any configuration interaction energy E_{conf} that includes second-order mixing, e.g., of ionic and neutral molecular states [4.38]:

$$U^{(2)} = E_{pol} + E_d + E_{conf} \quad . \tag{4.38}$$

If the atoms in the crystal are presumed to have atom-in-molecule polarizability tensors $\underline{\alpha}_i$ [4.42,43,64,72,80], then

$$E_{pol} = -\frac{1}{2} \frac{N}{Z} \sum_{i=1}^{M} \underline{F}_i^M \cdot \underline{\alpha}_i \cdot \underline{F}_i^M \quad , \tag{4.39}$$

where $\underline{F}_i^M$ is given by (4.24); if they are characterized only by a scalar average polarizability $\bar{\alpha}_i$, then E_{pol} becomes a scalar polarization energy E_{pol}^S [4.39,42,43, 48,55,59,64,72,80]:

$$E_{pol}^S = -\frac{1}{2} \frac{N}{Z} \sum_{i=1}^{M} \bar{\alpha}_i |\underline{F}_i^M|^2 \quad . \tag{4.40}$$

Also, the local dipoles μ_i^{hyb} can give rise to a dipolar field [4.48,85-88],

$$\underline{F}_i^D(\underline{r}_i) = \sum_{L=0}^{N/Z} \sum_{j=1}^{M} \left[\frac{3\mu_j \cdot (\underline{r}_i - \underline{r}_j - \underline{r}_L)(\underline{r}_i - \underline{r}_j - \underline{r}_L)}{|\underline{r}_i - \underline{r}_j - \underline{r}_L|^5} - \mu_j |\underline{r}_i - \underline{r}_j - \underline{r}_L|^{-3} \right] \quad , \tag{4.41}$$

which can also interact with either $\underline{\alpha}_i$ or $\bar{\alpha}_i$; however, this term has not yet been calculated. The London dispersion energy is given by

$$E_d = -\frac{3}{4} \frac{N}{Z} \sum_{i=1}^{M} \sum_{j=1}^{M'} \sum_{L=0}^{N/Z} \frac{\bar{\alpha}_i \bar{\alpha}_j}{|\underline{r}_i - \underline{r}_j - \underline{r}_L|^6} \frac{E_i E_j}{(E_i + E_j)} \quad , \tag{4.42}$$

where E_i may be taken as some measure of the average excitation energy of atom i. Usually, E_i is approximated by I_i, the ionization potential of the molecule to which atom i belongs [4.42,43]. If the geometric-mean law applies [4.21], then (4.42) simplifies to

$$E_d = -\frac{3}{8} \frac{N}{Z} \sum_{i=1}^{M} \sum_{j=1}^{M'} \sum_{L=0}^{N/Z} \frac{\bar{\alpha}_i \bar{\alpha}_j \sqrt{I_i I_j}}{|\underline{r}_i - \underline{r}_j - \underline{r}_L|^6} \quad . \tag{4.43}$$

This equation reduces then to a conveniently parametrizable form (see Chap.2),

$$E_d = -\frac{N}{2Z} \sum_{i=1}^{M} \sum_{j=1}^{M}{}' \sum_{L=0}^{N/Z} \frac{A_i A_j}{|r_i - r_j - r_L|^6} \quad ,$$ (4.44)

where A_i is either obtained from $(\sqrt{3}/2)\bar{\alpha}_i \sqrt{I_i}$ or is a parameter deduced from fitting model compound cohesive energies. The polarizability tensor $\underline{\alpha}_i$ has not yet been used in an equation patterned after (4.42).

As in the case of E_M, ϕ_i^M, F_{-i}^M, and E_{μ} (4.21,23,24,29), all intramolecular contributions must be subtracted from E_r, $F_{-i}^{D\mu}$, and E_d (4.31,32,41-44).

4.3 Born-Haber Cycles and Criteria for Ionicity

4.3.1 Born-Haber Cycles

The calculated cohesive energies U

$$U = U^{(0)} + U^{(2)} + \ldots \quad ,$$ (4.45)

should match fairly closely the experimental cohesive energy U_{exp}, which can be obtained from a suitably chosen Born-Haber cycle.

The simplest case is for the lattice that consists of essentially neutral molecules, i.e., with zero or little charge transfer from the electron donor (D) to the electron acceptor (A) (Fig.4.2).

The enthalpies of formation $\Delta H_f^0(D,c)$, $\Delta H_f^0(A,c)$, and $\Delta H_f^0(DA,c)$ are measurable by combustion calorimetry. The enthalpies of sublimation $\Delta H_s^0(D)$, $\Delta H_s^0(A)$ are obtainable from measurements of the temperature dependence of the vapor pressure. The experimental cohesive energy for a neutral lattice is

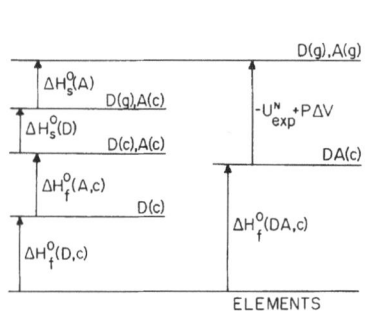

Fig. 4.2. Born-Haber cycle for a neutral DA crystal

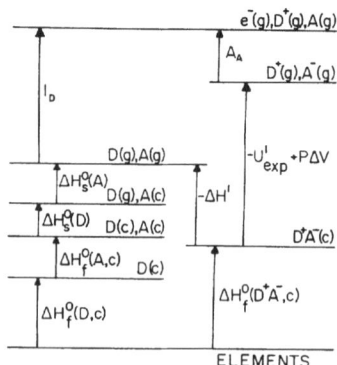

Fig. 4.3. Born-Haber cycle for a fully ionic D^+A^- crystal

$$U^N_{exp} = \Delta H^o_f(DA,c) - [\Delta H^o_f(D,c)+\Delta H^o_f(A,c)] - [\Delta H^o_s(D)+\Delta H^o_s(A)] - P\Delta V \quad . \tag{4.46}$$

In a unimolecular solid, e.g., anthracene, U^N_{exp} reduces to $-\Delta H^o_s - P\Delta V$. Here $-P\Delta V$ converts enthalpies into internal energies, i.e., corrects for finite temperature effects. By convention all U's are negative if the crystal is more stable than its constituents in the vapor phase.

For a fully ionic D^+A^- crystal (e.g., LiF) the relevant cohesive energy U^I_{exp} is given by (Fig.4.3)

$$U^I_{exp} = \Delta H^o_f(D^+A^-,c) - [\Delta H^o_f(D,c)+\Delta H^o_f(A,c)] - [\Delta H^o_s(D)+\Delta H^o_s(A)] - (I_D-A_A) - P\Delta V \quad , \tag{4.47}$$

where I_D is the vapor-phase adiabatic ionization potential of the donor

$$D \rightarrow D^+ + e^- \qquad \Delta H = I_D \quad , \tag{4.48}$$

measurable by mass spectrometry or other methods, and A_A is the vapor-phase adiabatic electron affinity of the acceptor

$$A + e^- \rightarrow A^- \qquad \Delta H = -A_A \quad . \tag{4.49}$$

The electron affinity of large organic molecules is difficult to determine experimentally. The best technique to date seems to be cesium beam electron attachment [4.89].

In order to compare $U^I_{exp} + (I_D-A_A)$ with purely thermochemical quantities, the ionic stabilization enthalpy ΔH^I is defined:

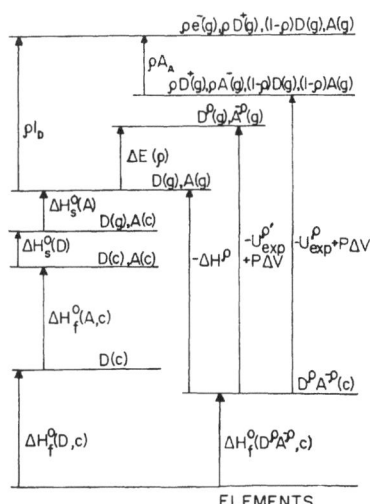

Fig. 4.4. Born-Haber cycle for a partially ionic $D^\rho A^{-\rho}$ crystal

$$\Delta H^I \equiv \Delta H^o_f(D^+A^-,c) - [\Delta H^o_f(D,c)+\Delta H^o_f(A,c)] - [\Delta H^o_s(D)+\Delta H^o_s(A)] \quad . \tag{4.50}$$

For a partially ionic crystal like $TTF^\rho TCNQ^{-\rho}$ the Born-Haber cycle (Fig.4.4) becomes more complicated. Assuming that the cost of ionizing the lattice to an extent $0 < \rho < 1$ is equivalent to the cost of ionizing a fraction ρ of the gaseous ions before forming the $D^\rho A^{-\rho}$ crystal [4.39], then the experimental cohesive energy is [4.35]

$$U^\rho_{exp} = \Delta H^o_f(D^{+\rho}A^{-\rho},c) - [\Delta H^o_f(D,c)+\Delta H^o_f(A,c)] - [\Delta H^o_s(D)+\Delta H^o_s(A)]$$
$$- \rho(I_D-A_A) - P\Delta V \quad . \tag{4.51}$$

This choice throws into the calculation of a theoretical U^ρ the onus of determining what polarization and other corrections must be added to the lattice energies discussed in Sects.4.2.2,3 to compensate for dielectric corrections for partial ionization in the solid state.

The other choice in Fig.4.4 is to define a fictitious energy level for a gas of partially ionized molecules $D^\rho(g)$ and $A^{-\rho}(g)$. The energy cost for forming these species from $D(g)$ and $A(g)$ is a new energy $\Delta E(\rho)$ which must reduce to 0 at $\rho = 0$ and to $I_D - A_A$ when $\rho = 1$. The difference between $\rho(I_D-A_A)$ and $\Delta E(\rho)$ has been first discussed as ΔE_{cov} in [4.39]. Recently SOOS [4.45] has studied $\Delta E(\rho)$ in detail, as will be discussed in Sect.4.3.2. If $\Delta E(\rho)$ is used, then the new cohesive energy is

$$U^{\rho'}_{exp} = \Delta H^o_f(D^{+\rho}A^{-\rho},c) - [\Delta H^o_f(D,c)+\Delta H^o_f(A,c)] - [\Delta H^o_s(D)+\Delta H^o_s(A)]$$
$$- \Delta E(\rho) - P\Delta V \quad . \tag{4.52}$$

Again, a thermochemically convenient reference quantity will be the partially ionic stabilization enthalpy ΔH^ρ

$$\Delta H^\rho \equiv \Delta H^o_f(D^{+\rho}A^{-\rho},c) - [\Delta H^o_f(D,c)+\Delta H^o_f(A,c)] - [\Delta H^o_s(D)+\Delta H^o_s(A)] \quad . \tag{4.53}$$

In summary, for partially ionic crystals U^ρ_{exp} requires that the theoretical lattice energy calculations contain the relevant solid state corrections so as to describe the process of partial ionization adequately. $U^{\rho'}_{exp}$ on the other hand requires that a valid description of the process of partial ionization be given on the single-molecule level, even though it is in reality a solid-state effect [4.45]. That is, the only difference between U^ρ_{exp} and $U^{\rho'}_{exp}$ is where one decides to put the difficult part of the calculation, namely the proper accounting for the process of partial ionization.

4.3.2 Criteria for Ionicity

It would be extremely convenient, both for the theoretician and for the experimentalist, to have some guidelines as to which of the various classical energies of Sect.4.2 (E_M, E_{cd}, E_μ, E_d, E_{pol}, E_r, E_k) is dominant in calculations of the theoretical cohesive energies U^N, U^I, U^o, $U^{o'}$. It would also be very useful if even from an approximate knowledge of the relevant Born-Haber cycle one could a posteriori explain, or a priori predict, the ionicity or degree of charge transfer ρ.

The traditional model ionic compound is sodium chloride (even though in NaCl ρ is probably close to 0.9). For NaCl U^I_{exp} = -764.0 kJ/mol [4.90], I_D = 496 kJ/mol [4.90], A_A = 348.6 kJ/mol [4.90], whereas the Madelung energy, assuming ρ = 1, is E_M = -861.04 kJ/mol [4.40,90]. Thus both U^I_{exp} and E_M are much greater in magnitude and opposite in sign with respect to $I_D - A_A$ = 147 kJ/mol. Furthermore, E_M is even greater than U^I_{exp}, so that the Madelung energy alone can stabilize the ionic lattice of sodium chloride.

Early experimental results on solid charge-transfer complexes indicated that they crystallized in what were later [4.24] called mixed simple regular (MSR) lattices, i.e., with DADADA overlap. These crystals could be further classified on the basis of their optical and paramagnetic spectra into "neutral" crystals and "ionic" crystals, i.e., the charge transfer ρ for the weak complexes (high I_D-A_A) was close to 0, and for the strong complexes (low I_D-A_A) ρ was fairly close to 1. These findings and a previous theoretical Hartree molecular-field model calculation [4.91] led McCONNELL, HOFFMAN, and METZGER (MHM) to propose [4.92] a simple criterion for when organic MSR crystals would be neutral (ρ=0) and when they would be ionic (ρ=1). Since E_M (4.21) scales as ρ^2 and $I_D - A_A$ should scale as ρ, the molecular-field energy $E^{(\rho)}_{MF}$,

$$E^{(\rho)}_{MF} = E_M\rho^2 + (I_D - A_A)\rho , \qquad (4.54)$$

should be useful at least in the vapor phase for predicting ρ. In particular, neutral crystals are predicted [4.92] if

$$E_M + I_D - A_A > 0 , \qquad (4.55)$$

and ionic (ρ=1) crystals are predicted [4.92] if

$$E_M + I_D - A_A < 0 , \qquad (4.56)$$

i.e., if the Madelung energy is sufficiently negative to offset the cost of ionization. A neutral-ionic interface could exist for crystals in which $E_M + I_D - A_A \sim 0$. In terms of the Born-Haber cycle (Fig.4.3), if E_M is close to U^I_{exp} then (4.56) is equivalent to requiring $\Delta H^I < 0$.

As has been reviewed elsewhere [4.41] and is repeated in Sect.4.4.2, (4.56) is satisfied for most ionic organic semiconductors. However, it fails for most organic metals [4.41]. To explain this, one must look more closely at (4.54).

If one calculates a neutral lattice Madelung energy E_M^N by using quantum-chemical charges [q_i^N, i=1,2,...,M] for the neutral molecules in (4.21), then E_M^N is reassuringly close to zero [4.39,42,43,44,55,61,62]. To calculate the Madelung energy for a partially ionic crystal, e.g., TTF$^{0.59}$TCNQ$^{-0.59}$, one may compute E_M for the fully ionic crystal using charges {q_i^I, i=1,2,...,M} in (4.21) and scale the result by ρ^2:

$$E_M(\rho) = \rho^2 E_M \quad . \tag{4.57}$$

Or, equivalently, one may compute the partially ionic Madelung energy E_M^ρ by using suitably interpolated charges q_i^ρ [4.39]:

$$q_i^\rho = (1-\rho)q_i^N + \rho q_i^I \qquad (i=1,2,...,M) \quad ; \tag{4.58}$$

this E_M^ρ is then very close to $\rho^2 E_M$. However, at all values of $\rho > 0$, $E_{MF}^{(\rho)}$ (4.54) is positive and has no minimum in the range $0 < \rho < 1$. That is, if $E_M < 0$ and $I_D - A_A > 0$, then a fractionally ionic state will never be stabilized [4.49] by (4.54). The Hartree molecular field arguments of [4.92] will lead to an energy maximum for $0 < \rho < 1$ and not to the desired minimum [4.49]. As is shown below, using E_k [4.49] and E_{cd}, E_μ, E_{pol} [4.42,43] in addition to E_M does not rectify the situation.

One possible remedy for this impasse is the use of the Wigner [4.93] lattice [4.39,42,43,46,56,58] at 0 K. For instance, one may imagine the $\rho = 0.5$ lattice "frozen" into a well-ordered superlattice that consists of fully charged ions interspersed with an equal number of neutral molecules so as to minimize repulsive Coulomb interactions. Thus a more negative E_M is obtained [4.39,42,43,46,56,58], but no dramatic improvements result if E_{cd}, E_μ, and E_{pol} are also calculated for the Wigner lattice [4.42]. Further, there is no physical evidence that the QODM exhibit such extreme ordering, and the Wigner lattice must be viewed as an idealized representation. Mixing $\rho = 1$ sites with $\rho = 0$ sites in Monte Carlo simulations of charge density waves in these partially ionic QODM may give a more realistic picture of the charge distribution, but the computer programs with such a modification of the lattice energy algorithms have not yet been written.

The other remedy for the inadequacies of (4.54) has been proposed by SOOS [4.45] and by BLOCH [4.94]. Both theories analyze carefully the process of partial ionization and reject the linear interpolation [4.92] $\rho(I_D - A_A)$ between $\rho = 0$ and $\rho = 1$ as too crude. Making appeal to arguments already used to explain mixed-valence inorganic complexes, SOOS [4.45] proposes that a smooth, quadratic fit,

$$\Delta E(\rho) = a_1\rho + a_2\rho^2 \quad , \tag{4.59}$$

to $I_D - A_A$ values at $\rho = -1$, 0, and +1 yields a better representation of the process of partial ionization than the linear interpolation scheme used in [4.39,92]. BLOCH, in a so far unpublished but more comprehensive cohesive energy theory uses the $\rho = 0$, 1, 2 states, and from a Hartree-Fock theory rather than a Hartree formalism [4.92] obtains a qualitatively similar description in the form of new chemical potentials $\Delta\mu_0$ and $\Delta\mu_1$ [4.94]. The use of (4.59) does indeed stabilize the partially charge-transferred state [4.45], as shown in Sect.4.4.4 below.

That is, by requiring that $\Delta E(\rho)$ have a quadratic or higher-order dependence on ρ and leaving $E_M(\rho)$ to its quadratic dependence on ρ, the intermediate charge trans-fer state can finally be stabilized.

This result [4.45] is particularly timely in that the semiconductor MSR salt TMPD TCNQ (when TMPD is N,N,N',N'-tetramethylparaphenylenediamine), a salt which was previously classified as fully ionic [4.24,92,95] has been shown by Raman spec-troscopy to have $\rho \approx 0.7$ (work cited in [4.45]). Therefore, it seems very important that the concept of ionicity and the detailed ρ-dependence of $\Delta E(\rho)$ be examined very carefully.

4.4 Lattice Energy Calculations

4.4.1 Lattice Energy Algorithms

Because of the conditional convergence of the Madelung sum, the Ewald fast-conver-gence double-series procedure [4.96] is used almost universally to obtain E_M (4.21). One study, however [4.53], uses the Evjen procedure [4.97]. Since Ewald's procedure is reviewed in standard solid state physics textbooks and is explicitly described in Chap.2, it will not be repeated here.

Ewald's series can be differentiated term by term to yield the Madelung electric field F_{-i}^M [4.39,42,43,48]. Ewald-type formulas also exist for the dipolar field F_{-i}^D [4.85-87]. Finally, Ewald's procedure has also been used to evaluate E_d, and ex-plicit expressions for E_d are also given in Chap.2.

The atom-in-molecule parameters (charge q_i, local hybrid moment μ_i^{hyb}, polariza-bility $\underline{\alpha}_i$, etc.) are obtained from ab initio quantum-chemical calculations or, more prosaically, from simple Hückel, extended Hückel, PPP (Pariser-Parr-Pople) theory, CNDO/2 (Complete Neglect of Differential Overlap), INDO (Intermediate Neglect of Differential Overlap) or MINDO/3 (Moderate Intermediate Neglect of Differential Overlap) [4.71] procedures.

4.4.2 Madelung Energy-Uniform Lattice

Before the results of the Madelung energy studies are reviewed, it is convenient to recall in Table 4.1 the principal types [4.24] of crystal lattice stackings found for organic donor-acceptor compounds.

In Table 4.2 are summarized the Madelung energies of two largely neutral and two mostly ionic MSR crystals, calculated assuming $\rho = 1$ [4.95]. The E_M values depend very little (up to 3%) upon the detailed choice of charge distribution [4.95] and hover around -400 kJ/mol. It is the magnitude of $I_D - A_A$ that decides whether (4.55) or (4.56) is obeyed. The more recent finding of incomplete charge transfer in TMPD TCNQ [4.45] is considered below in Sect.4.4.4. It should be noted that these organic MSR salts, with fairly good coordination of cations around each anion and vice versa,

Table 4.1. Types of lattice stacking in organic donor-acceptor (DA) crystals [4.24]

Lattice	Type	Ionicity ρ_{exp}	Type of linear stacks	Alteration of intermol. distances?	Examples
MSR	Mixed Simple Regular	0.0 to 1.0	DADA to $D^{+}A^{-}D^{+}A^{-}$	No	Hexamethylbenzene: para-chloranil, Naphthalene: TCNE, TMPD TCNQ, TMPD para-chloranil
SSR	Segregated Simple	1.0	$D^{+}D^{+}D^{+}D^{+}$	No	TMPD^{+}I^{-}, TMPD^{+}ClO$_4^{-}$ (T>186 K)
	Regular	1.0	$A^{-}A^{-}A^{-}A^{-}$	No	Rb^{+}TCNQ^{-} (II,T=300 K)
SSA	Segregated Simple	1.0	$D^{+}D^{+}D^{+}D^{+}$	Yes	TMPD^{+}ClO$_4$ (T<186 K)
	Alternating	1.0	$A^{-}A^{-}A^{-}A^{-}$	Yes	Na^{+}TCNQ^{-} (T=300 K), Rb^{+}TCNQ^{-} (I,T=110 K)
SCA	Segregated Complex	<1	$A^{-}A$ $A^{-}A$	Yes	Cs$_2$TCNQ$_3$, TEA TCNQ$_2$,
	Alternating	<1	$D^{+}D$ $D^{+}D$	Yes	(TMTSF)$_2$PF$_6$
SCR	Segregated Complex Regular	<1	$A^{-\rho}A^{-\rho}A^{-\rho}$ and/or $D^{-\rho}D^{-\rho}D^{-\rho}$	No	TMPD TCNQ$_2$, NMP TCNQ, TTF TCNQ

Table 4.2. Madelung energies E_M for four MSR crystals [4.41,95]

Crystal	Formal charge ρ_{exp}	E_M [kJ/mol][a]	$I_D - A_A$ [kJ/mol]	Classification by (4.54-56)
TMPD TCNQ	~1 ?	-380	333	Ionic (?)
TMPD para-chloranil	~1	-427	371	Ionic
Hexamethylbenzene: para-chloranil	~0	-426	540	Neutral
Naphthalene: Tetracyanoethylene	~0	-416	530	Neutral

[a] 1 eV = 96.487 kJ/mol. All energies are quoted per formula unit (DA pair).

have E_M values about half as large as for NaCl because of the lower charge per unit volume.

But even for the SSR and SSA lattices listed in Table 4.3 very large and negative E_M values are obtained. Where a great number of charge distributions was tested [4.36,40] the E_M values were again insensitive (to less than 3%) to choices between radically different atom-in-molecule charges q_i. For all these salts the nature of the counterion strongly suggests ρ_{exp} close to 1.0, and (4.56) is obeyed in every case. Although strong Coulomb repulsions occur along the stacking direction for the $D^+_\cdot D^+_\cdot$ (or the $A^-_\cdot A^-_\cdot$), these repulsions are surprisingly well compensated by attractions to the very small counterions. A very "dense" structure results, with E_M sometimes significantly more negative for SSA and SSR lattices (Table 4.3) than for the MSR lattices (Table 4.2). For NaTCNQ the enthalpy of formation has recently been measured [4.64,98]. The data are preliminary since about 20% of the sodium in NaTCNQ was converted to Na_2O_2 and corrections for this side reaction have not yet been carried out. Nevertheless, the resulting $U^I_{exp} = -690 \pm 20$ kJ/mol compares very well with $E_M = -510$ kJ/mol (Table 4.3) when a reasonable value for E_d is included: NaTCNQ is indeed a prevalently ionic solid. This calorimetric datum allays some previously expressed fears about binding energy defects in alkali and ammonium TCNQ salts [4.38].

Table 4.3. Madelung energies E_M for several SSR and SSA crystals

Crystal	Lattice type	E_M [kJ/mol]	I_D-A_A [kJ/mol]	Classification by (4.54-56)
$TMPD^+I^-$	SSR	-411 [4.36]	287	Ionic
$TMPD^+ClO_4^-$ (T=300 K)	SSR	-397 [4.40]	-	-
$TMPD^+ClO_4^-$ (T=110 K)	SSA	-399 [4.40]	-	-
Na^+TCNQ^- (T=300 K)	SSA	-510 [4.38]	227	Ionic
Rb^+TCNQ^- (I, T=110 K)	SSA	-454 [4.38] -457 [4.56]	133	Ionic
Rb^+TCNQ^- (II, T=300 K)	SSR	-448 [4.38]	133	Ionic

Table 4.4. Madelung energies E_M for two SCA crystals

Crystal	TCNQ charge from crystal structure $TCNQ_A$	$TCNQ_B$	E_M [kJ/mol]	I_D-A_A [kJ/mol]	Classification by (4.54-56)
Cs_2TCNQ_3	0	-1	-443^a [4.41]	213	Ionic
TEA $TCNQ_2$	$-\frac{1}{2}$	$-\frac{1}{2}$	-248 [4.41]	-	-

[a] E_M is quoted in kJ/(mol of Cs $TCNQ_{1.5}$) units.

Table 4.5. Madelung energies E_M at idealized full charge transfer ($\rho=1.0$) for metal-lic and semiconducting SCR crystals

Crystal	Formal charge per cation ρ_{exp}	QODM ?	E_M at $\rho=1.0$ [kJ/mol]	I_D-A_A [kJ/mol]	Ionicity predicted by (4.54-56)?
NMP TCNQ	0.94 [4.99] 0.33 [4.100] 0.66 [4.100]	yes	0 to 30 [4.37] 240 to -250 [4.58]	357?[a],[g]	No No
TTF TCNQ	0.59 [4.66]	yes	-187 to -225 [4.39] -148 to -239 [4.53] -170 to -210 [4.56]	389 [b],[g]	No No No
TTF Br$_{0.74}$	0.74 to 0.79 [4.101]	yes	-376 [4.46] -307 to -329 [4.46][i]	310 [b],[h]	Yes
TTT$_2$I$_3$	0.50	yes	-96 [4.60][j]	-	-
TMPD TCNQ$_2$	0.50	no	-271 [4.41]	333 [c],[g]	No
HMTTF TCNQ HMTTF TCNQF$_4$ HMTSF TCNQ	0.72 [4.102] 1.00 [4.112] 0.28 [4.103] 0.74 [4.104]	yes no yes	-253 [4.51] -251 [4.112] -249 [4.63]	347 [d],[g] >347 [d],[g]	No No No
HMTSF TCNQF$_4$	0.58 [4.103] 1.00 [4.105]	no	-273 [4.63]	-	No
HMTSF TNAP	-	yes	-20 [4.63]	-	No
TMTSF TCNQ red	0.13 [4.103]	no	-297 [4.63]	365 [e],[g]	No
TMTSF TCNQ black	0.70 [4.103] 0.57 [4.106]	yes	-29 [4.63]	365 [e],[g]	No
TMTTF TCNQ	0.48 [4.103]	yes	-39 [4.63]	349 [f],[g]	No

[a] Assuming I_D(NMP.) = 6.4 eV from indirect considerations [4.58];

[b] Assuming I_D(TTF) = 6.83 eV [4.107]; [c] Assuming I_D(TMPD) = 6.25 eV [4.108];

[d] Assuming I_D(HMTTF) = 6.4 eV [4.51]; [e] Assuming I_D(TMTSF) = 6.58 eV [4.109];

[f] Assuming I_D(TMTTF) = 6.42 eV [4.109]; [g] Assuming A_A(TCNQ) = 2.8 eV [4.89];

[h] Assuming A_A(Br) = 3.56 eV [4.110]; [i] Wigner lattice calculation for $\rho = 0.75$;

[j] Uniform lattice calculation for $\rho = 0.5$. No Ewald fast convergence methods were used.

The Madelung energy data for two SCA crystals are listed in abbreviated form in Table 4.4. Some computer tests on the charge distribution for these salts are described in detail in [4.41].

The E_M calculations for many SCR crystals are given in Table 4.5. Except where noted, only the $\rho = 1.0$ uniform lattice data are tabulated. Most of the crystals are metallic. For all, except the TTF Br$_x$ studies, (4.56) is *not* satisfied. For many of them (NMP TCNQ, HMTSF TNAP, black form of TMTSF TCNQ, and TMTTF TCNQ) E_M is close to zero. For all of them E_M varies unduly with charge model. For this reason, varying $\{q_i, i=1,2,...,M\}$ until one is found that makes E_M change from +240 to -250 kJ/mol [4.58] does not seem to be a valid procedure. As stated before [4.39] in these QODM SCR structures the crystal seems to want to minimize something other than

$E_M + I_D - A_A$. Thus, a binding energy "defect" [4.39] occurs throughout Table 4.5. For TTF TCNQ this defect persists at all ρ, $0 < \rho < 2$ [4.39]. For comparison with E_M, an experimental cohesive energy $U^I_{exp} = -635 \pm 16 - P\Delta V$ kJ/mol at $\rho = 1.0$ ($U^\rho_{exp} = -471 \pm 16 - P\Delta V$ kJ/mol at $\rho = 0.59$) has been determined [4.35]. Thus, the E_M values of Table 4.5 are only 24% of U^I_{exp}. The same thermochemical data [4.35] yield $\Delta H^\rho = -235 \pm 6$ kJ/mol [cf. (4.53)].

It is quite obvious from Table 4.5 that one must look beyond the simple uniform-lattice Madelung energy to stabilize the SCR lattices.

4.4.3 Madelung Energy-Wigner Lattice

As was mentioned in Sect.4.3.2, the Wigner lattice model has the obvious advantage of minimizing, for a given ρ, the Coulomb repulsions along segregated stacks [4.39, 42,46,56,58,111].

Several Wigner lattice E_M calculations have been carried out; they are summarized in Table 4.6. For the SSA compound Rb^+TCNQ^-(I) the expected ρ_{exp} is 1.0, and indeed, assuming a $\rho = 0.5$ Wigner lattice does not bring any energetic advantage: $E_M(\rho=1.0)$ can by itself stabilize the lattice (Table 4.3).

The QODM crystal TTF $Br_{0.74}$ has an incommensurate Br^- sublattice, which forces the assumption of some form of the Wigner lattice before E_M calculations can be carried out. By varying the occupation of the Br^- sublattice, several $E_M(\rho)$ calculations when coupled with $\rho(I_D-A_A)$ yielded a very nice minimum of $E_M(\rho) + \rho(I_D-A_A)$ at a value of $\rho \approx 0.75$ very close to ρ_{exp}. Also, thanks to the small size of the bromide counterion, the minimum was negative (-88 kJ/mol). That is, the MHM criterion [4.92] coupled with a Wigner crystal calculation provides a minimum at $\rho = 3/4$ for TTF $Br_{0.74}$ [4.46].

For the QODM TTT_2I_3, the incommensurate I_3^- sublattice can be translated in a Wigner lattice model and provides reasonable minima in E_M which occur at the crystallographically determined "best" iodine atom positions and yet are shallow enough to provide I_3^- "sliding" [4.60].

The difficult cases are TTF TCNQ and NMP TCNQ. For both salts, assuming $\rho = 0.5$, there is a 60% to 150% "improvement" in $E_M(\rho)$ as one passes from the uniform lattice to the Wigner lattice. However, for TTF TCNQ one is still very far from U^ρ_{exp}. For both NMP TCNQ and TTF TCNQ the MHM criteria (not very applicable to NMP TCNQ) yield no minima at intermediate ρ unless one assumes that I_D-A_A is depressed (at $\rho=1.0$) by about 1 eV of polarization energy [4.111]. For NMP TCNQ such an analysis is weakened by the fact that the neutral radical NMP· is not a thermodynamically stable species, but in the absence of thermochemical data for NMP TCNQ there seems to be only this choice. Worse yet, the considerable dependence of E_M on details of the charge model for TTF TCNQ becomes an extreme case in NMP TCNQ (Table 4.5). Allowing the Madelung site potential to affect directly the NDO quantum-chemical calculation until self-consistency is achieved [4.47,57,58] may add apparent elegance to the

Table 4.6. Madelung energies E_M of SCR and SSA lattices: Wigner lattices compared to uniform lattices (using the same charge models)

Crystal	Lattice type	ρ_{exp}	Uniform lattice $E_M(\rho_{exp})$ [kJ/mol]	Wigner lattice $E_M(\rho)$ [kJ/(mol of formula unit)]					
				$\rho=1/5$	$\rho=1/4$	$\rho=1/3$	$\rho=1/2$	$\rho=2/3$	$\rho=3/4$
TTF TCNQ	SCR	0.59 [4.66]	-93 [4.39]	---	-77 [4.111] -78 [4.46]	-106 [4.111]	-154 [4.111] -151[a][4.39] -154[b][4.39] -130 [4.46]	-174 [4.111]	-183 [4.111]
NMP TCNQ	SCR	0.33 [4.100] 0.66 [4.100] 0.94 [4.99]	-232[e] [4.58] -222[c,e][4.58]	-48 [4.58]	-68 [4.58]	-93 [4.58]	-149 [4.58] -149[c][4.58]	-174 [4.58]	-181 [4.58]
Rb[+]TCNQ[-] (I, T=110 K)	SSA	1.0	-457[d] [4.56] -454[d] [4.38]	---	---	---	-190 [4.58] -214 [4.46]	---	---
TTF Br$_{074}$	SCR	0.74 to 0.79 [4.101]	---	-79 to -93 [4.46]	-101 to -117 [4.46]	-132 to -153 [4.46]	-207 to -226 [4.46]	-279 to -297 [4.46]	-307[d] to -329 [4.46]
TTT$_2$I$_3$	SCR	0.5	-96[f] [4.60]	---	---	---	-230[f] to -260[f][4.60]	---	---

[a] Wigner lattice A [4.39]; [b] Wigner lattice B [4.39]; [c] Atom charges forced to self-consistency with Madelung site potentials [4.41,58]; [d] Location of energy minimum of $E_M(\rho)+\rho(I_D-A_A)$; [e] Assuming $\rho=1.0$; [f] No Ewald fast convergence used.

resulting "self-consistent" E_M values, but hard evidence has yet to be obtained for this dramatic redistribution of the intramolecular charge distribution.

4.4.4 Beyond the Madelung Energy

Considerable efforts have been made to find for NMP TCNQ and TTF TCNQ other cohesive energy contributions that would fit the experimental Born-Haber cycle data (for TTF TCNQ [4.35]) and, more generally, stabilize the partially ionic lattice.

For narrow-band metals such as TTF TCNQ and TTF $Br_{0.74}$, it was of no surprise to find that E_k(TTF TCNQ) = -24 kJ/mol [4.39], E_k(TTF $Br_{0.74}$) = -10 kJ/mol [4.46] were small.

By a conventional packing energy analysis (Chap.2 and [4.1-22]) a value for E_d (4.44) and E_r (4.32) have been obtained [4.61,62] for TTF TCNQ: $E_d + E_r$ = -225 kJ/mol (Table 4.1, col.3 of [4.61]). This, together with a scalar average of all E_M calculations (!) evaluated at $\rho = 0.59$, E_M^{ρ} = -73 kJ/mol [4.61], yielded a calculated U^{ρ} = -298 kJ/mol [*not* -757 kJ/mol as in Table 6 of [4.61]: the reference state at infinity for $E_M(\rho)$ calculations is *not* D(g), A(g), but rather $\rho D^+(g)$, $\rho A^-(g)$, $(1-\rho)D(g)$, $(1-\rho)A(g)$]. This U^{ρ} value is closer to U_{exp}^{ρ} = -484 kJ/mol - PΔV [4.35] or to U_{exp}^{ρ} = -511 kJ/mol - PΔV [4.61] than the simple E_M value alone, but is still far from accounting for all the binding energy in TTF TCNQ. Nonetheless, this calculation has brought into focus the importance of dispersion forces in binding organic ionic crystals and has sparked a fruitful new avenue of research (Chap.5).

It is possible that a significant portion of what packing energy programs call the van der Waals or dispersion energy E_d (4.44) may in fact "lump" other contributions, e.g., the polarization energy.

Thus in a conformational energy calculation of neutral TTF [4.55], E_d = -149.6 kJ/mol, E_M = -0.4 kJ/mol, E_{pol}^S = -162 to -175 kJ/mol, and E_r = 62.3 kJ/mol. These values agree well with E_d = -156 kJ/mol, E_r = 60 kJ/mol, E_M = 0 to -7 kJ/mol as obtained by [4.61,62] and indicate that E_{pol}^S is of the same order of magnitude as E_d!

Because E_{pol}^S promised to be so large, a considerable amount of effort was devoted to modify NDO quantum-chemical programs to produce reliable atom-in-molecule polarizability tensors $\underline{\alpha}_i$ along with the customary charges q_i and local hybrid moments μ_i^{hyb} [4.41,64,72,80]. The MINDO/3 procedure, suitably modified to perform finite perturbation (FP) calculations [4.80], yields for TTF, TCNQ, TMPD and their radical ions acceptable polarizabilities $\underline{\alpha}_i$ [4.72]. Figure 4.5 illustrates the molecular $\underline{\alpha}$ for TTF and TTF^+ and the atom-in-molecule q_i, $|\mu_i^{hyb}|$, and $\bar{\alpha}_i$ for TTF^+. The polarizabilities depend strongly on molecular geometry and increase considerably as one passes from the neutral molecule to the radical ion, but are severely underestimated in the π-π overlap direction [4.72].

Nevertheless, these new atom-in-molecule parameters were included in a study of cohesive energies of TTF TCNQ (Table 4.7) and TMPD TCNQ (Table 4.8) as a function of uniform charge transfer and also for the $\rho = 1/2$ TTF TCNQ Wigner lattice.

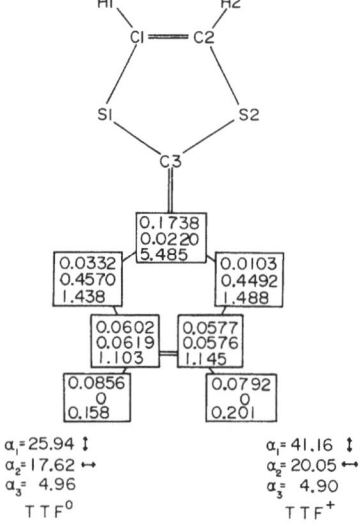

$\alpha_1 = 25.94 \updownarrow$
$\alpha_2 = 17.62 \leftrightarrow$
$\alpha_3 = 4.96$

TTF⁰

$\alpha_1 = 41.16 \updownarrow$
$\alpha_2 = 20.05 \leftrightarrow$
$\alpha_3 = 4.90$

TTF⁺

Fig. 4.5. MINDO/3-FP charges q_i, hybrid moments $|\mu_i^{hyb}|$, and scalar average polarizabilities for TTF⁺ in the TTF TCNQ crystal structure. In each box the top entry is q_i in |electrons|, the second is $|\mu_i^{hyb}|$ in |electron| Å, the bottom entry is $\bar{\alpha}_i$ in Å³. Also given at the bottom of the figure are the three principal axis values of the molecular polarizability tensor $\underline{\alpha}$ for TTF⁰ (left) and TTF⁺ (right)

Table 4.7. Cohesive energy contributions [kJ/mol] for TTF TCNQ using MINDO/3-FP q_i, $\bar{\alpha}_i$, $\underline{\alpha}_i$, and μ_i^{hyb} data from [4.42]

	$\rho = 0.0$ Uniform lattice	$\rho = 0.5$ Uniform lattice	$\rho = 0.5$ Wigner lattice A	$\rho = 1.0$ Uniform lattice
E_M	-2	-52	-135	-194
E_{pol}^S	-33	-52	-63	-123
E_{pol}	-70	-126	-155	-286
E_{cd}	90	146	145	206
E_μ	-12	-10	13	-8
$E_4 = E_M + E_{pol} + E_{cd} + E_\mu$	6	-42	-132	-282
E_d	-266	---	---	-391

Table 4.8. Cohesive energy contributions [kJ/mol] for TMPD TCNQ using MINDO/3-FP q_i, $\bar{\alpha}_i$, $\underline{\alpha}_i$, and μ_i^{hyb} data from [4.43]

	$\rho = 0$ Uniform lattice	$\rho = 1.0$ Uniform lattice
E_M	-5	-317
E_{pol}^S	-24	-104
E_{pol}	-53	-217
E_{cd}	47	77
E_μ	1	2
$E_4 = E_M + E_{pol} + E_{cd} + E_\mu$	-10	-455

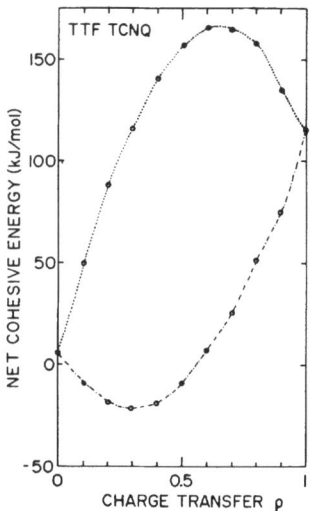

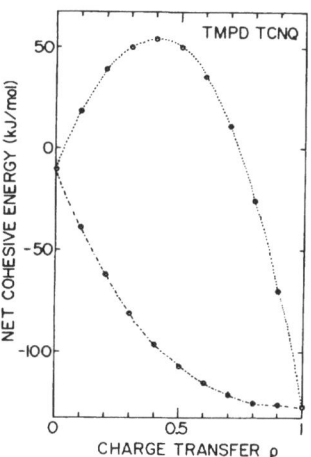

Fig. 4.6. Comparison of the MHM linear interpolation scheme $E_4+\rho(I_D-A_A)(\ldots)$ with the Soos $E_4+\Delta E(\rho)$ scheme $(-.-.-)$ for the crystal binding energy of TTF TCNQ

Fig. 4.7. Comparison of the MHM linear interpolation scheme $E_4+\rho(I_D-A_A)(\ldots)$ with the Soos $E_4+\Delta E(\rho)$ scheme $(-.-.-)$ for the crystal binding energy of TMPD TCNQ

The surprising feature of Table 4.7 is that the heretofore unsuspected charge-dipole term is so large as to cancel E_M at $\rho = 1$. Thus for TTF TCNQ the zeroth-order cohesive energy $E_M + E_{cd} + E_\mu$ is almost zero (the situation in Table 4.5 for NMP TCNQ, HMTSF TNAP, black TMTSF TCNQ and TMTTF TCNQ!), while for TMPD TCNQ, E_M is much larger than E_{cd} and this cancellation does not occur. The E_d values obtained in Table 4.7 using (4.43) compare well with those of [4.61] and are charge-transfer dependent.

The E_4 values in Table 4.7 plus $E_d + E_r = -225$ kJ/mol [4.61] (or some other ρ-dependent estimate of E_d+E_r) should be comparable to U^ρ_{exp} for TTF TCNQ, but are in fact too small. The dominating attractive Madelung energy contributions in the $\rho = 0.5$ Wigner crystal limiting model help yield $E_4 = -132$ kJ/mol. This is still far short of $U^{0.50}_{exp} = -435$ kJ/mol.

Maybe improved E_{pol} calculations (using polarization functions to obtain better $\underline{\alpha}$ values), and better $E_d + E_r$ estimates will ultimately close the binding energy gap.

It was hoped that the use of linear interpolation formulas similar to (4.58) for μ^{hyb}_i and $\underline{\alpha}_i$ for the $\rho \neq 1.0$ uniform lattice model would yield E_{pol} that would be cubic in ρ, thus breaking the tyranny of the MHM expression (4.54), and somehow establishing a cohesive energy minimum at intermediate ρ. This hope did not materialize [4.42,43]. An alternate hope, that the use of $E_4 = E_M + E_{pol} + E_{cd} + E_\mu$ in place of E_M together with Soos' quadratic interpolation expression $\Delta E(\rho)$ (4.59) would yield a minimum of $E_4 + \Delta E(\rho)$ at a ρ value closer to the experimental one for TMPD TCNQ and TTF TCNQ, was similarly dashed. In Figs.4.6,7 the MHM scheme yields maxima as usual at intermediate ρ for $E_4 + \rho(I_D - A_A)$. In Fig.4.6, the Soos minimum for $E_4 +$ $\Delta E(\rho)$ occurs at $\rho \sim 0.3$, lower than the experimental $\rho_{exp} = 0.59$ for TTF TCNQ. In Fig.4.7 the $E_4 + \Delta E(\rho)$ minimum occurs at too high a ρ value in comparison with $\rho_{exp} \sim$ 0.7 for TMPD TCNQ.

In conclusion, the calculations of cohesive energy contributions beyond the Madelung energy have yielded large and charge-transfer-dependent polarization energies and a surprisingly significant antibinding E_{cd} term. However, the location of the energy minimum still occurs at the wrong ρ in Soos' scheme. Clearly much good work still lies ahead.

Acknowledgements. It is a pleasure to thank Prof. Heimo J. Keller (Universität Heidelberg) and Dr. Pierre Delhaes (Université de Bordeaux I — Centre Paul Pascal) for their kind and generous hospitality. The support of the National Science Foundation (grants DMR-77-09314 and 78-16998) is gratefully acknowledged.

References

4.1 A.I. Kitaigorodsky: *Organic Chemical Crystallography* (Consultants Bureau, New York 1961)

4.2 A.I. Kitaigorodsky: *Molecular Crystals and Molecules* (Academic Press, New York 1973)

4.3 A.I. Kitaigorodsky, K.V. Mirskaya: Sov. Phys. Crystallogr. *6*, 408 (1962)

4.4 K.V. Mirskaya: Sov. Phys. Crystallogr. *8*, 167 (1963)

4.5 A.I. Kitaigorodsky, K.V. Mirskaya: Sov. Phys. Crystallogr. *9*, 137 (1964)

4.6 A.I. Kitaigorodsky, K.V. Mirskaya: Sov. Phys. Crystallogr. *10*, 121 (1965)

4.7 N.A. Ahmed, A.I. Kitaigorodsky, K.V. Mirskaya: Acta Crystallogr. *B27*, 867 (1971)

4.8 A.I. Kitaigorodsky, K.V. Mirskaya: Mater. Res. Bull. *7*, 1271 (1972)

4.9 K.V. Mirskaya, V.V. Nauchitel': Sov. Phys. Crystallogr. *17*, 56 (1972)

4.10 V.V. Nauchitel', K.V. Mirskaya: Sov. Phys. Crystallogr. *16*, 891 (1972)

4.11 K.V. Mirskaya, I.E. Kozlova, V.F. Bereznitskaya: Phys. Status Solidi B *62*, 291 (1974)

4.12 K.V. Mirsky: Acta Crystallogr. *A32*, 199 (1976)

4.13 D.E. Williams: J. Chem. Phys. *45*, 3770 (1966)

4.14 D.E. Williams: J. Chem. Phys. *47*, 4680 (1967)

4.15 D.E. Williams: Acta Crystallogr. *A25*, 464 (1969)

4.16 D.E. Williams: Trans. Am. Crystallogr. Assoc. *6*, 21 (1970)

4.17 D.E. Williams: Acta Crystallogr. *A27*, 452 (1971)

4.18 D.E. Williams: Acta Crystallogr. *A28*, 629 (1972)

4.19 D.E. Williams: Acta Crystallogr. *A30*, 71 (1974)

4.20 D. Hall, D.E. Williams: Acta Crystallogr. *A31*, 56 (1975)

4.21 D.E. Williams, T.L. Starr: Comput. Chem. *1*, 173 (1977)

4.22 T.L. Starr, D.E. Williams: Acta Crystallogr. *A33*, 771 (1977)

4.23 F.H. Herbstein: *Crystalline pi-Molecular Compounds*, in Perspectives in Structural Chemistry, Vol.4, ed. by J.D. Dunitz, J.A. Ibers (John Wiley & Sons, New York 1972)

4.24 Z.G. Soos: Annu. Rev. Phys. Chem. *25*, 121 (1974)

4.25 A.F. Garito, A.J. Heeger: Acc. Chem. Res. *7*, 232 (1974)

4.26 H.J. Keller (ed.): Low-Dimensional Cooperative Phenomena, NATO Advanced Study Institutes, Ser.B, Vol.7, Lectures presented at Starnberg, West Germany 1974 (Plenum Press, New York 1975)

4.27 H.J. Keller (ed.): *Chemistry and Physics of One-Dimensional Metals*, NATO Advanced Study Institutes, Ser.B, Vol.25, Lectures presented at Bolzano, Italy 1976 (Plenum Press, New York 1977)

4.28 L. Pàl, G. Grüner, A. Jànossy, J. Sòlyom (eds.): *Organic Conductors and Semiconductors*, Proc. of the Int. Conf., Siòfok, Hungary 1976, Lecture Notes in Physics, Vol.65 (Springer, Berlin, Heidelberg, New York 1977)

4.29 J.S. Miller, A.J. Epstein (eds.): *Synthesis and Properties of Low-Dimensional Materials*, Proc. of the New York Academy Conf., 1977, Ann. N.Y. Acad. Sci. *313* (1978)

4.30 J.T. Devreese, R.P. Evrard, V.E. van Doren (eds.): *Highly Conducting One-Dimensional Solids* (Plenum Press, New York 1979)

4.31 S. Barišić, A. Bjeliš, J.R. Cooper, B. Leontić (eds.): *Quasi-One-Dimensional Conductors I, II*, Proc. of the Int. Conf., Dubrovnik, Yugoslavia 1978, Lecture Notes in Physics, Vols.95.96 (Springer, Berlin, Heidelberg, New York 1979)

4.32 W.E. Hatfield (ed.): *Molecular Metals*, NATO Conf. Proc., Les Arcs, France 1978 (Plenum Press, New York 1979)

4.33 L. Alcàcer (ed.): *Physics and Chemistry of Low-Dimensional Solids*, NATO Advanced Study Institutes, Vol.C56, Lectures presented at Tomar, Portugal 1979 (D. Reidel, Dodrecht, Holland 1980)

4.34 R.H. Boyd: J. Chem. Phys. *38*, 2529 (1963)

4.35 R.M. Metzger: J. Chem. Phys. *66*, 2525 (1977)

4.36 R.M. Metzger: J. Chem. Phys. *57*, 1870 (1972)

4.37 R.M. Metzger: J. Chem. Phys. *57*, 2218 (1972)

4.38 R.M. Metzger: J. Chem. Phys. *63*, 5090 (1975)

4.39 R.M. Metzger, A.N. Bloch: J. Chem. Phys. *63*, 5098 (1975)

4.40 R.M. Metzger: J. Chem. Phys. *64*, 2069 (1976)

4.41 R.M. Metzger: Ann. N.Y. Acad. Sci. *313*, 145 (1978)

4.42 R.M. Metzger: *The Cohesive Energy of TTF TCNQ as a Function of Charge Transfer*, J. Chem. Phys., in press

4.43 R.M. Metzger, F.M. Wiygul, Z.G. Soos: *The Cohesive Energy of TMPD TCNQ as a Function of Charge Transfer*, to be published

4.44 Z.G. Soos, A.J. Silverstein: Mol. Phys. *23*, 755 (1972)

4.45 Z.G. Soos: Chem. Phys. Lett. *63*, 179 (1979)

4.46 J.B. Torrance, B.D. Silverman: Phys. Rev. *B15*, 788 (1977)

4.47 W.D. Grobman, B.D. Silverman: Solid State Commun. *19*, 319 (1976)

4.48 B.D. Silverman, W.D. Grobman, J.B. Torrance: Chem. Phys. Lett. *50*, 152 (1977)

4.49 B.D. Silverman: Phys. Rev. *B16*, 5153 (1977)

4.50 B.D. Silverman: Phys. Rev. *B17*, 2482 (1978)

4.51 B.D. Silverman, S.J. LaPlaca: J. Chem. Phys. *69*, 2585 (1978)

4.52 B.D. Silverman: J. Chem. Phys. *71*, 3592 (1979)

4.53 A.J. Epstein, N.O. Lipari, D.J. Sandman, P. Nielsen: Phys. Rev. *B13*, 1569 (1976)

4.54 A.J. Epstein, N.O. Lipari, P. Nielsen, D.J. Sandman: Phys. Rev. Lett. *34*, 914 (1975)

4.55 D.J. Sandman, A.J. Epstein, J.S. Chickos, J. Ketchum, J.S. Fu, H.A. Scheraga: J. Chem. Phys. *70*, 305 (1979)

4.56 V.E. Klymenko, V.Ya. Krivnov, A.A. Ovchinnikov, I.I. Ukrainsky, A. Shvets: Zh. Eksp. Teor. Fiz. *69*, 240 (1975) [Engl. transl.: Sov. Phys.-JETP *42*, 123 (1975)]

4.57 V.E. Klymenko, I.I. Ukrainsky: Zh. Strukt. Khim. *19*, 412 (1978)

4.58 V.E. Klymenko, V.Ya. Krivnov, A.A. Ovchinnikov, I.I. Ukrainsky: J. Phys. Chem. Solids *39*, 359 (1978)

4.59 A.A. Ovchinnikov, V.Ya. Krivnov, V.E. Klymenko, I.I. Ukrainsky: *On the Charge Distribution and the Lattice Distortion of Quasi-One-Dimensional Systems*, in [4.28] p.103

4.60 K. Kamaràs, M. Kertèsz: Solid State Commun. *28*, 607 (1978)

4.61 H.A.J. Govers: Acta Crystallogr. *A34*, 960 (1978)

4.62 H.A.J. Govers, C.G. De Kruif: Acta Crystallogr. *A36*, 428 (1980)

4.63 A.N. Bloch, R.M. Metzger, F.M. Wiygul: to be published

4.64 R.M. Metzger: *Cohesive Energy and Ionicity*, in [4.33] p.233

4.65 G.A. Thomas, D.E. Schafer, F. Wudl, P.M. Horn, D. Rimai, J.W. Cook, D.A. Glocker, M.J. Skove, C.W. Chu, R.P. Groff, J.L. Gillson, R.C. Wheland, L.R. Melby, M.G. Salamon, R.A. Craven, G. De Pasquali, A.N. Bloch, D.O. Cowan, V.V. Walatka, R.E. Pyle, R. Gemmer, T.O. Poehler, G.R. Johnson, M.G. Miles, J.D. Wilson, J.P. Ferraris, T.F. Finnegan, R.J. Warmack, V.F. Raaen, D. Jerome: Phys. Rev. *B13*, 5105 (1976)

4.66 F. Denoyer, R. Comès, A.F. Garito, A.J. Heeger: Phys. Rev. Lett. *35*, 445 (1975)
4.67 D. Jerome, A. Mazaud, M. Ribault, K. Bechgaard: J. Phys. Lett. *41*, L85 (1980)
4.68 P.-O. Löwdin: Phil. Mag. Suppl. *5*, 1 (1956)
4.69 F.E. Harris, L. Kumar, H.J. Monkhorst: Int. J. Quantum Chem. *5*, 527 (1971)
4.70 V.L. Moruzzi, A.R. Williams, J.F. Janak: Phys. Rev. *B15*, 2854 (1977)
4.71 G. Klopman, R.C. Evans: *"The Neglect of Differential Overlap Methods of Mole-cular Orbital Theory"*, in Semiempirical Methods of Electronic Structure Cal-culation. Part A: Techniques, ed. by G.A. Segal (Plenum Press, New York 1977)
4.72 R.M. Metzger: J. Chem. Phys. *74*, 3458 (1981)
4.73 CNDO/2 is Version 2 of the Complete Neglect of Differential Overlap semiempirical molecular orbital theory; cf. [4.71] and J.A. Pople, D.L. Beveridge: *Approximate Molecular Orbital Theory* (McGraw Hill, New York 1970)
4.74 S.F. O'Shea, D.P. Santry: Theor. Chim. Acta *37*, 1 (1975)
4.75 R.M. Metzger: J. Chem. Phys. *57*, 4847 (1972)
4.76 P.-O. Löwdin: J. Chem. Phys. *18*, 365 (1950)
4.77 R. Landshoff: Z. Phys. *102*, 201 (1936)
4.78 J.D. Jackson: *Classical Electrodynamics* (Wiley, New York 1962) Chap.4
4.79 S.P. McGlynn, L.G. Vanquickenborne, M. Kinoshita, D.G. Carroll: *Introduction to Applied Quantum Chemistry* (Holt, Rinehart and Winston, New York 1972)
4.80 R.M. Metzger: J. Chem. Phys. *74*, 3444 (1981)
4.81 I. Stakgold: *Boundary-Value Problems of Mathematical Physics*, Vol.II (Macmillan, New York 1968) p.111
4.82 J. Kommandeur: private communication
4.83 A. Fröman, P.-O. Löwdin: J. Phys. Chem. Solids *23*, 75 (1962)
4.84 W.C. Nieuwpoort, G. Blasse: J. Inorg. Nucl. Chem. *30*, 1635 (1968)
4.85 M.H. Cohen, F. Keffer: Phys. Rev. *99*, 1128 (1955)
4.86 R.L. Bush: Phys. Rev. *B12*, 5698 (1975)
4.87 D.A. Dunmur: Mol. Phys. *23*, 109 (1972)
4.88 P.G. Cummins, D.A. Dunmur, R.W. Munn, R.J. Newham: Acta Crystallogr. *A32*, 847 (1976)
4.89 R.N. Compton, C.D. Cooper: J. Chem. Phys. *66*, 4325 (1977)
4.90 C. Kittel: *Introduction to Solid State Physics*, 4th ed. (Wiley, New York 1971)
4.91 J.I. Krugler, C.G. Montgomery, H.M. McConnell: J. Chem. Phys. *41*, 2421 (1964)
4.92 H.M. McConnell, B.M. Hoffman, R.M. Metzger: Proc. Natl. Acad. Sci. USA *53*, 46 (1965)
4.93 E. Wigner: Trans. Faraday Soc. *34*, 678 (1938)
4.94 A.N. Bloch: Bull. Am. Phys. Soc. *25*, 255 (1980)
4.95 R.M. Metzger: J. Chem. Phys. *57*, 1876 (1972)
4.96 P.P. Ewald: Ann. Phys. (Leipzig) *64*, 253 (1921)
4.97 E. Evjen: Phys. Rev. *39*, 675 (1932)
4.98 E.S. Arafat, C.S. Kuo, R.M. Metzger: to be published
4.99 M.A. Butler, F. Wudl, Z.G. Soos: Phys. Rev. *B12*, 4708 (1975)
4.100 J.P. Pouget, S. Megtert, R. Comès, A.J. Epstein: Phys. Rev. *B21*, 486 (1980)
4.101 S.J. LaPlaca, P.W.R. Corfield, R. Thomas, B.A. Scott: Solid State Commun. *17*, 635 (1975)
4.102 R. Comès, G. Shirane: quoted in [4.51]
4.103 T.J. Kistenmacher: "Partial Charge Transfer and Charge Density Wave Modulation in the TTF-TCNQ Family of Quasi-One-Dimensional Organic Materials", in *Modu-lated Structures — 1979*, Kailua Kona, Hawai, AIP Conf. (Am. Institute of Physics, New York 1979)
4.104 A.N. Bloch: unpublished results
4.105 A.N. Bloch, T. Cape, R.P. van Duyne: unpublished
4.106 J.P. Pouget: private communication
4.107 R. Gleiter, E. Schmidt, D.O. Cowan, J.P. Ferraris: J. Electron. Spectrosc. *2*, 207 (1973)
4.108 M. Batley, L.E. Lyons: Mol. Cryst. *3*, 357 (1968)
4.109 R. Gleiter, M. Kobayashi, J. Spanget-Larsen, J.P. Ferraris, A.N. Bloch, K. Bechgaard, D.O. Cowan: Ber. Bunsenges. Phys. Chem. *79*, 1218 (1975)
4.110 L. Pauling: *The Nature of the Chemical Bond*, 3rd ed. (Cornell University Press, Ithaca, NY 1960)

4.111 I.I. Ukrainsky, V.E. Klymenko, A.A. Ovchinnikov: Electrostatic energy and electronic structure of donor-acceptor molecular crystals based on tetra-cyanoquinodimethan. Preprint ITP-75-89E, Institute of Theoretical Physics, Ukrainian Acad. Sci. Kiev, USSR (1975)
4.112 J.B. Torrance, J.J. Mayerle, K. Bechgaard, B.D. Silverman, Y. Tomkiewicz: Phys. Rev. *B22*, 4960 (1980)

Additional Reference with Title

Metzger, R.M.: Madelung Energy and Theoretical Zero-Field Splittings of (2:1)-(Tetra-methyltetraselenafulvalenium hexafluorophosphate), $(TMTSF)_2PF_6$. J. Chem. Phys. *75*, 482-484 (1981)

5. Slipped Versus Eclipsed Stacking of Tetrathiafulvalene (TTF) and Tetracyanoquinodimethane (TCNQ) Dimers

B. D. Silverman

With 25 Figures

The quasi-one-dimensional organic metal TTF-TCNQ exhibits ordering of like molecules (TTF or TCNQ, and their radical ions) in segregated stacks. In order to understand this ordering, calculations are performed for the isolated dimers (TTF$_2$, TCNQ$_2$, and their ions) as a function of "slip". The dimer calculations first use semiempirical molecular orbital (MO) theory (extended Hückel, CNDO/2) and finally the Gordon-Kim density functional method.

Because intermolecular dispersion forces are neglected by the minimum-basis MO theories, these yield less than satisfactory dimer geometries. The Gordon-Kim procedure is far more successful in predicting the correct minimum-energy geometries. This implies that dispersion forces are extremely important in understanding the crystal structure and ultimately the transport properties of TTF-TCNQ.

5.1 Introductory Comments

There has been considerable interest in the electrical properties of organic salts composed of donor and acceptor molecules [5.1-3]. Salts such as TTF-TCNQ (tetrathiafulvalene-tetracyanoquinodimethan) and NMP-TCNQ (N-methylphenazinium-tetracyanoquinodimethanide) have electrical conductivities of approximately 10^2 $(\Omega\ cm)^{-1}$ or higher and exhibit a whole host of interesting properties associates with quasi-one-dimensional electrical behavior. The most significant structural feature of these materials is the presence of segregated stacks or face-to-face stacking of like planar molecules along a given direction of the crystal. This is the direction that provides an easy path along which conduction electrons can travel in the crystal. The details of the stacking of molecules along this direction are, therefore, of interest. In particular, the intermolecular separation between molecules on a stack is important in determining the magnitude of the one-dimensional electronic bandwidth. Also, the degree of slip of adjacent molecules on a stack is not only important in determining the magnitude of this bandwidth but can also be important in determining the sign of the band dispersion.

One expects that the origin of the observed molecular geometry in organic salts will be difficult to determine in detail. First, the phenomenological procedures [5.4] that have been relatively successful in predicting neutral molecular structures have not been extended and used in connection with organic salts. Second, molecular-orbital programs developed in connection with problems involving strong chemical bonds exhibit deficiencies when used to examine the interactions between molecules at or near van der Waals separations. On the other hand, since the structure of conducting organic compounds is of importance in determining the electronic properties of these materials, investigations of the origin of certain specific geometric details exhibited by a compound such as TTF-TCNQ and related materials have been initiated with presently available computational procedures.

The purpose of the present chapter is to review the results of calculations aimed at understanding the stability and local geometry of TCNQ and TTF molecules in TTF-TCNQ and in certain other related materials. The materials discussed will be those shedding light on the interactions primarily responsible for the particular geometric relationship between adjacent molecules in a segregated stack. A good deal of discussion will be devoted to examining the relative orientation of two adjacent TTF molecules in neutral crystalline TTF as well as in TTF-TCNQ. The "TTF story" is an interesting one and provides a lesson in the pitfalls of using simple molecular-orbital programs to determine the relative orientation between molecules at van der Waals separations. The story is certainly not concluded. To a large extent, the present chapter represents the author's attempt to understand why the simple molecular-orbital calculations, e.g., extended-Hückel and CNDO/2 procedures, yield such an incomplete picture of bonding in an organic salt such as TTF-TCNQ and what one might do to make the picture more complete.

5.2 Geometry of Donor-Acceptor π Complexes: Slipped versus Eclipsed Stacking

It is common for large planar molecules to form structures composed of molecules packed face to face [5.5-7]. Organic salts composed of donor and acceptor molecules generally exhibit stacking in which the donor and acceptor molecules alternate in position along a stack. One expects such alternating structure to result primarily from the electrostatic attraction between the donor and acceptor molecules. A much less common stacking mode is the one in which like molecules stack face to face along a given direction of the crystal. Such stacking has been commonly designated "segregated stacking". In an organic salt, one expects this type of stacking to result from the delocalization of electrons along the stack or, in other words, from the one-dimensional electronic banding. The electronic conduction band in the organic donor-acceptor compounds arises from the π-orbital overlap of nearest-neighbor electronically equivalent donors (TTF) or acceptors (TCNQ) on a stack. Segregated stacking in donor-acceptor compounds is, therefore, essential for the high values of electrical conductivity achieved in donor-acceptor compounds such as TTF-TCNQ. Segregated stacks of either TTF or TCNQ molecules fall into one of two categories, namely stacks with a slipped (Fig.5.1a,2a) or eclipsed (Fig.5.1b,2b) geometry. TCNQ and TTF stacking in TTF-TCNQ is of the slipped kind for which adjacent molecules on a stack are displaced relative to one another along the long molecular axis [5.8,9]. When viewed normal to the molecular plane, adjacent molecules exhibit the so-called "ring-to-bond" coordination. Projections of adjacent TTF and TCNQ molecules normal to the molecular plane are shown in Fig.5.1a,2a, respectively. An example of eclipsed stacking for which the molecules are directly over each other is provided by the non-stoichiometric TTF-halides [5.10,11]. Eclipsed stacking of dimers has also been found in the stochiometric 1:1 TTF-halide salts [5.10,11]. Nearly eclipsed TCNQ stacking has been found in the high-temperature phases of Na-TCNQ and K-TCNQ [5.12,13]. In general, there are slight deviations from an exactly eclipsed stacking geometry and a variation in the degree of relative displacement between molecules exhibiting slipped stacking. However, the stacking geometry observed in a wide variety of TCNQ compounds and the smaller number of known TTF compounds generally falls into one or the other of these two categories.

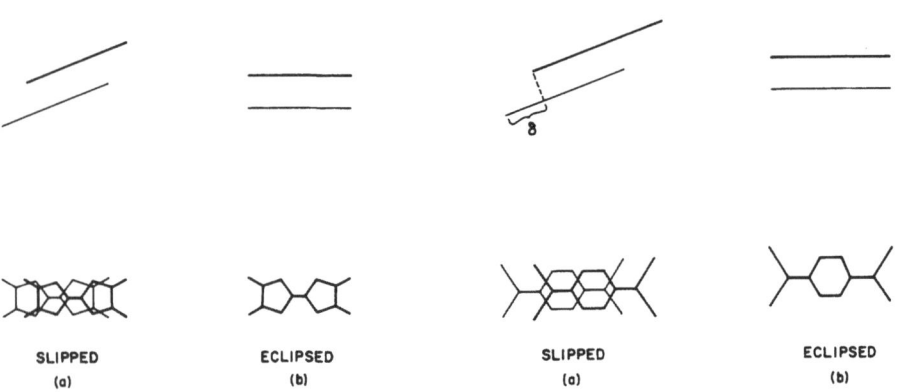

| SLIPPED | ECLIPSED | SLIPPED | ECLIPSED |
| (a) | (b) | (a) | (b) |

Fig. 5.1a,b. Segregated stacking of TTF (tetrathiafulvalene): (a) slipped stacking, (b) eclipsed stacking

Fig. 5.2a,b. Segregated stacking of TCNQ (tetracyanoquinodimethane): (a) slipped stacking, (b) eclipsed stacking

The particular type of stacking geometry holds important consequences for electronic banding [5.14] since the π-orbital bandwidth depends upon the overlap of adjacent molecules on a stack. For example, at fixed intermolecular separation, eclipsed stacking will always yield the largest electronic bandwidth. This is so since the orbital overlap is optimized for this geometry. Also, eclipsed stacking always yields an inverted one-dimensional π-orbital band since the lowest energy at the zone edge results from the build-up of electron density midway between the molecular planes.

It is well known [5.14] that the TCNQ band in TTF-TCNQ is noninverted as a result of the slipped geometry of the TCNQ stack. The band dispersion, i.e., inverted as opposed to noninverted, can therefore depend upon the type of stacking. Differences in stacking, therefore, lead to important differences in the properties of the one-dimensional electronic bands and hence motivate one to understand the origin of the observed stacking geometries.

It certainly would be of interest to understand why given types of molecules form segregated as opposed to nonsegregated or mixed stacks. Up to the present, however, no calculations have been performed for which the energies of segregated stack structures and mixed stack structures have been compared. Since one expects the Coulomb interaction between donor and acceptor molecules to be primarily re-sponsible for the alternating stack structures of charge transferred compounds, this interaction must be treated with reasonable accuracy. On the other hand, cal-culations have been performed with respect to the simpler problem of the relative stability of eclipsed versus slipped segregated stacking. The insight gained as a result of tackling the simpler problem of slipped versus eclipsed stacking should assist in understanding why segregated stacks in the donor-acceptor compounds form in the first place.

5.3 Molecular-Orbital Calculations

5.3.1 Extended-Hückel Calculations

Molecular-orbital calculations were initially performed by CHESNUT and MOSELEY [5.15] in their attempt to understand the local conformation or stacking geometry of TCNQ molecules in different molecular crystals. They used an extended-Hückel procedure which only treated the π-orbital framework. The total energy of the TCNQ dimer was obtained for different orientations of the monomers. Such a procedure is useful in determining the relative stability of different molecular geometries that are due to covalency effects, namely, effects due to the overlapping of partially filled π or-bitals. This is an important, but incomplete part of the story. Perhaps a useful way of viewing the contributions to the interaction energy between molecules near van der Waals separations in an organic salt is the following. Holding the geometry of the organic salt fixed, let us reverse the charge transfer until the molecules in the crystal become neutral. Let us call the interaction energy between molecules of this neutral aggregate E_N. Now, let us slowly turn on the charge transfer until it establishes its final value in the crystal. The interaction energy will change by an amount which we will call ΔE_{CT}, the change in the total molecular interaction en-ergy due to the charge transfer. The total dimer energy E_T can therefore be written as

$$E_T = E_N + \Delta E_{CT} \quad .$$ (5.1)

Molecular-orbital programs, such as extended-Hückel, CNDO/2, MINDO/3, etc., purport to calculate E_T, but in reality yield only a reasonable estimate of ΔE_{CT}. E_N is poorly described by these molecular-orbital programs. First, all Hartree-Fock cal- culations neglect the van der Waals interaction or electronic correlation between molecules, which is an essential ingredient of E_N. Furthermore, the nature of the approximations of molecular-orbital programs such as extended-Hückel, CNDO/2, etc., namely truncated basis, neglect of differential overlap, particular empirical param- etrization, etc., do not yield the accurate behavior of the kinetic energy with in- termolecular separation or orientation. The variation in kinetic energy with vary- ing intermolecular separation plays a fundamental role [5.16] in establishing this separation. If, on the other hand, one uses these molecular-orbital programs to investigate conformations involving strong chemical bonds at chemical bond distances, this is a more reasonable procedure since it is ΔE_{CT} which essentially determines the equilibrium bond lengths and angles. For molecules interacting at or near van der Waals distances, however, the contributions from E_N as well as from ΔE_{CT} are important in determining the equilibrium geometry. It is important to keep this in mind when discussing the extended-Hückel and CNDO/2 results that will be presented. Later on we will discuss in some detail the more accurate determination of E_N.

CHESNUT and MOSELEY performed calculations for dimers consisting of donor and acceptor molecules as well as for dimers composed of two TCNQ monomers, appropriate to segregated stack structures. We will restrict our discussion to the latter case. Figure 5.3 shows the energy contour maps determined by CHESNUT and MOSELEY for the neutral, singly charged, and doubly charged TCNQ dimer. The value of the energy at any point on one of the contours is associated with a dimer geometry such that one molecule is at the position given by the reference figure and the other molecule has its center at the point on the contour. The molecular axes of both molecules composing the dimer are held parallel. The calculations were performed for a mole- cular geometry and intermolecular separation near the values found in the NMP-TCNQ and Cs_2TCNQ_3 organic salts. The calculated energy minima are indicated by the squares in Fig.5.3. Experimentally observed projections are designated by a triangle, circle, and star which are, respectively, the observed projections given in Fig.5.4a-c. Some general trends of the well depths associated with the energy minima are apparent as the net charge on the dimer is varied. One sees that as the charge on the dimer is increased, or as one goes from the neutral to doubly-charged dimer, the well depth for the slipped geometry gets deeper. The singly-charged dimer has a secondary min- imum for a conformation that is near the overlap observed between the two molecules, each lacking a crystallographic center of symmetry, on a TCNQ segregated stack in Cs_2TCNQ_3. For a larger value of charge transfer, the doubly-charged dimer, the low- est energy minimum is found near a geometry for which the molecules are directly over each other, the eclipsed geometry. It is of interest to note that all three

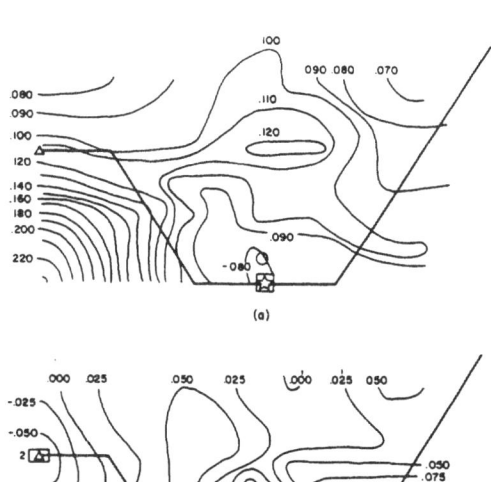

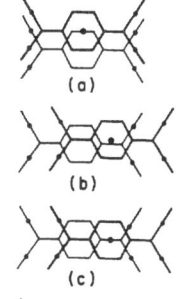

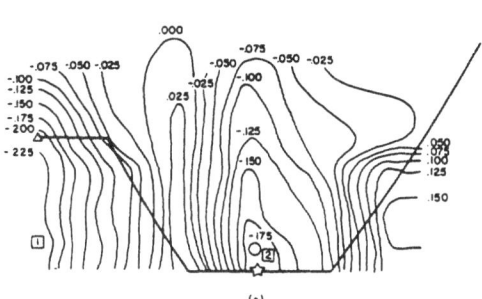

Fig. 5.4a-c. Line drawings are given showing the experimentally observed projections for (a) two adjacent TCNQ molecules on a stack in Cs_2TCNQ_3, each lacking a crystallographic center of symmetry; (b) two adjacent molecules on a stack in Cs_2TCNQ_3, one with and one without a crystallographic center of symmetry; (c) two adjacent molecules on a stack in NMP-TCNQ. From [5.15]

◄ **Fig. 5.3a-c.** Energy contour maps for (a) the uncharged species, (b) the anion, and (c) the dianion of $(TCNQ)_2$. The experimental projections found in Cs_2TCNQ_3 are denoted by a triangle (noncentric-noncentric case) and by a circle (centric-noncentric case). That found in NMP-TCNQ is denoted by a star. Calculated minima are shown by numbered squares. From [5.15]

minima predicted by the extended-Hückel calculation have been observed in one or another of the TCNQ segregated stack structures. Not only do the TCNQ molecules in NMP-TCNQ and the centric-noncentric TCNQ pair of molecules in Cs_2TCNQ_3 exhibit the slipped geometry, but every good organic conductor ($\gtrsim 1 \, \Omega^{-1} cm^{-1}$) investigated to date that consists of TCNQ segregated stacks exhibits the slipped geometry. Also, a nearly eclipsed conformation, predicted by the calculation for the doubly charged TCNQ dimer, has been found for certain TCNQ salts that are apparently completely charge transferred, namely, one electronic charge per monomer. These are the alkali TCNQ salts for which a number of structural studies have been performed [5.12]. It should finally be noted that the well-depth minimum calculated for the neutral dimer is at positive energy as are all other values on the energy contours and hence the calculation predicts no bonding for the neutral complex.

The CHESNUT and MOSELEY work was performed prior to all of the interest and excitement generated by studies [5.17] of TTF-TCNQ. Subsequent to the discovery of TTF-TCNQ, molecular-orbital calculations on TCNQ and TTF dimers were carried out by

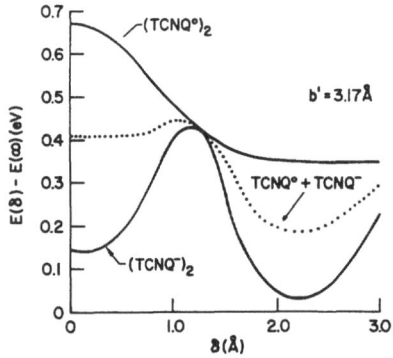

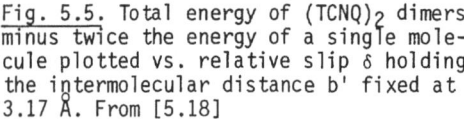

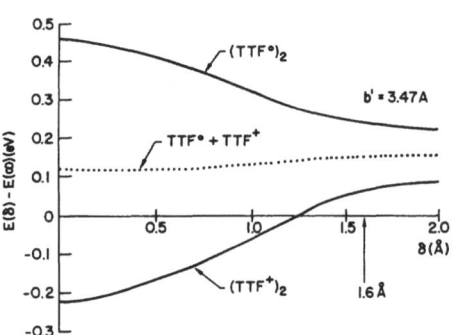

Fig. 5.5. Total energy of (TCNQ)$_2$ dimers minus twice the energy of a single molecule plotted vs. relative slip δ holding the intermolecular distance b' fixed at 3.17 Å. From [5.18]

Fig. 5.6. Total energy of (TTF)$_2$ dimers minus twice the energy of a single molecule plotted vs. relative slip δ holding the intermolecular distance b' fixed at 3.50 Å. From [5.18]

BERLINSKY and coworkers [5.18]. They calculated the total energy of TTF as well as TCNQ dimers. Instead of the π-orbital-only calculation utilized by CHESNUT and MOSELEY, they used the more general extended-Hückel calculation [5.19], which treats the σ as well as the π orbitals. Monomer displacements along the long molecular axis as well as variations in intermolecular separation were treated. The internal molecular geometries were taken from X-ray diffraction studies [5.8] of TTF-TCNQ. Since variations in the total energy as a function of intermolecular separation yield inconclusive results, our discussion at present will neglect this. Figure 5.5 is a plot of normalized total TCNQ dimer energy as a function of slipping one of the monomers with respect to the other along the long molecular axis. Results are shown for the neutral, singly charged and doubly charged dimer. The intermolecular separation is held fixed at the value observed in TTF-TCNQ. The results are essentially in agreement with the earlier π-only Hückel calculations. The singly and doubly charged dimers exhibit energy minima at an eclipsed as well as a slipped conformation. Also, as the electronic charge on the dimer is increased from one to two, the well depth at the eclipsed geometry is lowered with respect to the well depth at the slipped geometry. As previously mentioned, this is consistent with the observation that whereas the TCNQ stack geometry in TTF-TCNQ is slipped, more highly charge-transferred salts such as the alkali-TCNQ salts exhibit a nearly eclipsed geometry. There are apparent differences, however, between the π-only and the extended-Hückel calculations. One should note that the ordering of the well depths at the two minima is reversed for the doubly-charged dimer. One expects that this particular ordering, as well as the one we will obtain when we discuss the CNDO/2 results, is not significant because of the approximate nature of the molecular-orbital calculations. The extended-Hückel neutral dimer also exhibits no energy minimum, in agreement with the π-orbital-only result.

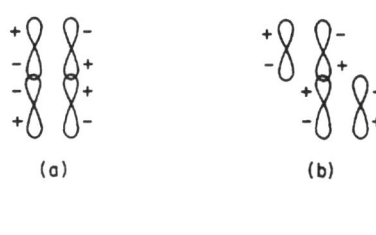

Fig. 5.7a-d. Hypothetical p orbital monomer wave functions in simulation of dimer overlap. (a) Eclipsed geometry — even monomer wave function; (b) slipped geometry — even monomer wave function; (c) eclipsed geometry — odd monomer wave function; (d) slipped geometry — odd monomer wave function

The extended Hückel results for the TTF dimer are significantly different from the TCNQ results. Total energy as a function of slip is shown in Fig.5.6. For TTF, the singly and doubly ionized dimers exhibit only one minimum in the total energy. This minimum is at the eclipsed geometry. Since TTF stacking in TTF-TCNQ is slipped, this prompted BERLINSKY and coworkers to state that TTF stacks the way that it does as a result of the presence of TCNQ stacks.

This difference between the number of energy minima as a function of slip for the TTF and TCNQ dimers can be understood to result simply from the difference in symmetry between the TCNQ affinity level and the TTF ionization level. In Fig.5.7, we have used a hypothetical p-orbital monomer wave function to simulate the π orbital of the dimer in order to illustrate this point. The lowest unoccupied orbital (LUMO) of neutral TCNQ has even symmetry and the π orbital shown in Figs.5.7a,b has such symmetry. Figure 5.7a shows a dimer with the eclipsed geometry for which the monomer orbitals are out of phase. In Fig.5.7b, we show that it is energetically favorable for the monomer orbitals to change phase as one monomer is slipped with respect to the other. It should be remembered that the bonding molecular orbital optimizes the buildup of electronic charge between monomers. The highest occupied orbital (HOMO) of neutral TTF has odd symmetry and the phases of the two p orbitals of our schematic π-orbital monomer in Fig.5.7c,d have been chosen to have this symmetry. Figure 5.7c shows the dimer in the eclipsed geometry with the two monomer orbitals out of phase. In Fig.5.7d we show that as one monomer is slipped with respect to the other, it does not pay energetically for the monomer molecular-orbitals to reverse phase·with respect to each other. This is the reason the ionized TTF dimers exhibit a single minimum in total energy over the regime of slip investigated.

One sees from Fig.5.6 that as the charge transfer or state of ionization of the dimer is increased, the well depth of the minimum at the eclipsed geometry becomes deeper. Later on, in connection with our discussion of the CNDO/2 results, we will point out that such behavior is consistent with the known structure of a small number of TTF compounds, namely, the greater the charge transfer, the greater the tendency to eclipsed stacking.

The results for the neutral TTF dimer show no energy minimum over the range of slip investigated. This behavior is similar to that obtained for the neutral TCNQ dimer. A recent calculation [5.20] has shown that an energy minimum in the total energy as a function of slip for the neutral TTF dimer can be obtained at a value of slip of approximately 2.0 Å. This extended-Hückel calculation has been performed for an intermolecular separation of 3.47 Å. The presence of this minimum in the total energy requires the absence of 3d orbitals on the sulfur atoms.

BERLINSKY and coworkers also calculated the total dimer energy of TCNQ and TTF as a function of intermolecular separation. The relative value of slip was held at the observed values. As previously mentioned, the results were inconclusive since no minima in the total energy were found near the observed intermolecular separations in the crystal. This is symptomatic of the difficulties encountered when using only Hartree-Fock results to determine the interactions between molecules at or near van der Waals separations. At these separations, it is also necessary to include correlation or, in other words, the van der Waals energy.

5.3.2 CNDO/2 Calculations

Hückel calculations neglect certain very large interactions. Within the context of semiempirical procedures such as CNDO/2 [5.21] these interactions can be labeled core-core, core-electron, and electron-electron. The implicit assumption that such interactions largely cancel or are unimportant with respect to certain chemical properties is, therefore fundamental to the successful use of any Hückel calculation. It has been argued [5.22] that systems with a net electronic charge are inappropriate for investigation with a Hückel procedure. CHESNUT and MOSELEY expressed some concern about the modification of their Hückel results if core-core interactions were included. Generally speaking, one might expect that inclusion of the electrostatic interactions between the monomers of the charged dimer would lead to the stabilization of slipped structures as opposed to eclipsed structures, since for eclipsed structures one has like charge distributions in proximity.

It is, therefore, of interest to perform molecular-orbital calculations on TCNQ and TTF dimers that include, at least in some approximation, the large interaction energies that are not explicitly treated by the extended-Hückel procedure. Such calculations have recently been performed [5.23] with use of the CNDO/2 molecular-orbital program [5.21]. This program enables one to obtain the total energy for a fixed dimer geometry and also allows one to identify explicitly the core-core contribution to this energy as well as the electronic contribution. These two contributions make up the total energy. The electronic energy is composed of a core-electron and an electron-electron contribution. The electron-electron part is in turn made up of a direct Coulomb (Hartree) part plus the exchange part. The core-electron interaction includes the Hückel off-diagonal resonance contributions as well as certain large additional diagonal electrostatic contributions. All of these interactions are

treated in the approximate manner that is appropriate to CNDO/2. A detailed discussion of these approximations has been given by POPLE and BEVERIDGE in [5.21].

Calculations have been performed on TCNQ and TTF dimers with molecular geometries
taken from the TTF-TCNQ crystal structure [5.8]. As found from the extended-Hückel
calculations, the variation of intermolecular separation at some fixed value of slip
does not lead to meaningful results and hence will not be discussed. We will discuss
the results of calculations performed as a function of slip along the long molecular
axis at the value of the intermolecular separation observed in the crystal. In Fig.
5.8 we show the results of calculation for the singly charged TCNQ dimer. The two
contributions to the total energy, the core-core and electron interaction energies
are plotted as a function of slip, δ, along the long molecular axis of TCNQ. The
sum of these two contributions, the total energy, is also given. It is of interest
to note that the core-core repulsive interaction energy decreases rapidly as the
molecules are slipped away from the eclipsed geometry. This had been surmised by
CHESNUT and MOSELEY. The curves shown in Fig.5.8 have been placed at arbitrary values on the figure, since the actual values vary widely, and we are mainly interested in comparing the changes in energies as a function of slip. Even though the
core-core interaction energy decreases, one sees an increase in electronic energy
as the molecules are slipped away from the eclipsed geometry. Over the range of slip
investigated, this increase in electronic energy closely follows the decrease in the
core-core energy so that variations of hundreds of electron volts in the individual
quantities result in only variations of tenths of volts in their sum, the total energy. The behavior of the total energy looks very much like the result obtained from
the extended-Hückel calculation. There are two energy minima, one at the eclipsed
geometry and the other near the slipped geometry found for TCNQ in TTF-TCNQ. It
should be emphasized that the CNDO/2 total TCNQ dimer energy is roughly twenty times
the extended-Hückel value. This is essentially due to the large electrostatic interactions neglected in the extended-Hückel procedure. Variations in these energies,
therefore, largely cancel as a function of slip, and as a result, do not appear to
be important in determining the equilibrium dimer geometry. We will show later on
that electronic correlation and/or the van der Waals (dispersion) energy does contribute significantly to the determination of the equilibrium geometry of the neutral
dimer. One, therefore, expects that electron-electron contributions are important
with respect to such determination in the organic salts, but not in the Hartree or
mean-field manner treated by the CNDO/2 program.

Total energy is plotted as a function of slip for neutral, singly and doubly
charged TCNQ dimers in Fig.5.9. Both singly and doubly charged dimers exhibit minima
at the eclipsed geometry and near the slipped geometry observed in the crystal. In
contrast to the extended-Hückel results, the energy is lowest at the eclipsed geometry. This could well reflect a general tendency of zero differential overlap
schemes to yield lower energies for the more closely connected structures [5.24].
In any event, one does not expect either the extended-Hückel or the CNDO/2 procedures

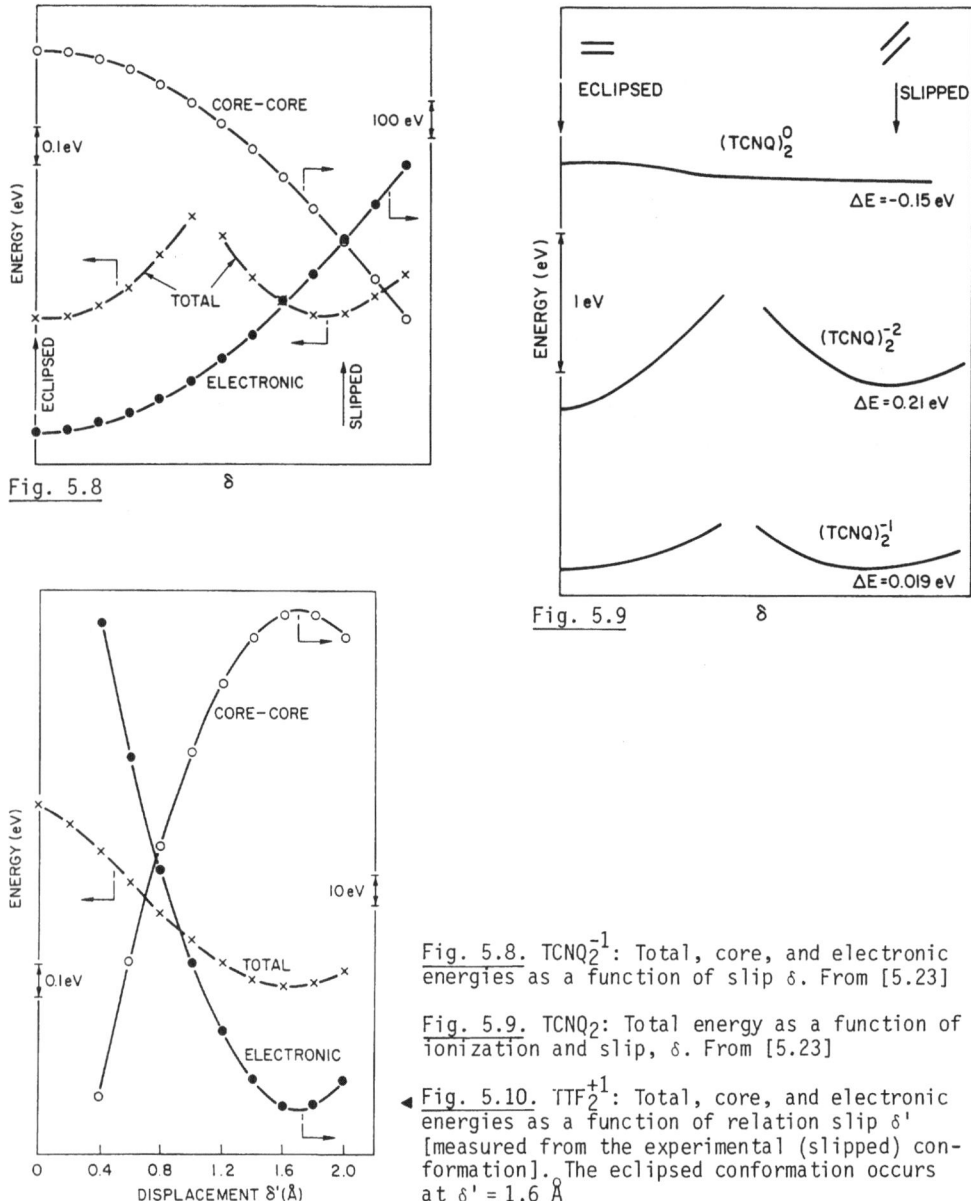

Fig. 5.8. $TCNQ_2^{-1}$: Total, core, and electronic energies as a function of slip δ. From [5.23]

Fig. 5.9. $TCNQ_2$: Total energy as a function of ionization and slip, δ. From [5.23]

◄ Fig. 5.10. TTF_2^{+1}: Total, core, and electronic energies as a function of relation slip δ' [measured from the experimental (slipped) conformation]. The eclipsed conformation occurs at $\delta' = 1.6$ Å

to yield accurately the small differences in energy between the energy minima. From Fig.5.9 one further notes that the energy at the eclipsed minimum is lowered with respect to the energy at the slipped minimum, as the charge transfer or net electronic charge on the TCNQ dimer is increased. This is consistent with the observation that fully charge-transferred donor-acceptor compounds such as the alkali TCNQ salts [5.12,13] exhibit nearly eclipsed TCNQ stacking, whereas all of the known incompletely charge-transferred compounds exhibit slipped stacking. On the other hand, not all of the fully charge-transferred compounds exhibit the nearly eclipsed geometry. HMTTF-TCNQF$_4$ [5.25], HMTSF-TCNQF$_4$ [5.26], and Rb(TCNQ)-II [5.27], which are believed to be fully charge-transferred, exhibit slipped TCNQ stacking.

Examination of the monomer affinity level splitting upon formation of the neutral dimer is of interest. For small electronic overlap between monomers, twice this splitting is the electronic bandwidth associated with the TCNQ stack [5.28]. It is found that this bandwidth goes through maxima at the eclipsed and slipped geometries found for the charged dimers. The filling of the banded affinity levels or conduction band in TCNQ is therefore important in establishing the observed stacking geometry. The band is inverted for the eclipsed geometry. The wave function associated with the lowest energy at the Brillouin zone edge consists of monomer wave functions of neighboring molecules on a stack that are out of phase. At the slipped geometry, the wave function associated with the lowest energy at the Brillouin zone center consists of monomer wave functions that are in phase.

As we pointed out in connection with the extended-Hückel calculations, the results for the TTF dimer are significantly different. This is also true of the CNDO/2 results. In Fig.5.10 we show the core-core and electronic energy as well as their sum, the total energy as a function of slip, δ. These results are for the singly ionized dimer. Again we note that whereas the core-core and electronic energies vary by hundreds of volts, close cancellation between these two quantities yields variations in the total energy of only tenths of volts over the range of slip investigated. We see, as found for the extended-Hückel calculation, that there is only one minimum in the total energy, and it is found at the eclipsed conformation. [It should be noted that, for this particular figure only, displacements have been measured from the observed slipped conformation [5.8]: a value of $\delta' = 1.6$ Å of the abscissa therefore indicates an eclipsed conformation]. A similar result is found for the doubly-charged dimer. One also finds that the energy minimum for the doubly charged dimer is lower than the minimum for the singly charged dimer. One might then conclude that the more highly charge-transferred compounds of TTF should exhibit a tendency for the TTF molecules to stack with an eclipsed geometry. LA PLACA [5.29] has pointed out that among the small number of known TTF charge transfer compounds, those having the higher value of charge transfer do exhibit an eclipsed geometry. The TTF stacks in TTF-TCNQ as well as neutral crystalline TTF are slipped. For the nonstoichiometric halides with charge transfer of approximately 0.72 and the 1:1 salts consisting of dimers, the stacking is eclipsed [5.10,11]. It will be of interest to see if such a relationship between stacking geometry and the charge transfer is exhibited by other TTF salts. These arguments, however, might not be appropriate for certain other TTF-donor-like molecules. HMTTF-TCNQF$_4$ is believed to be fully charge transferred. However, the HMTTF stacking is slipped. The methylene groups might well further stabilize the slipped geometry so that even at full charge transfer, the stacking remains slipped. Also, interactions between molecules on different stacks should contribute to the stabilization of one stacking geometry over another. It is important to recognize that the TTF structures exhibiting the eclipsed geometry involve a small anion with a spherically symmetric charge distribution.

The total energy of the neutral TTF dimer has an energy minimum at the eclipsed stacking geometry. This result differs from the extended-Hückel result (Fig.5.6) [5.18], which shows no minimum over the investigated range of slip. The origin of this difference has been discussed recently [5.20]. Both extended-Hückel and CNDO/2 procedures are, however, inappropriate for treating interactions between neutral, closed-shell monomers. Of significance, however is the single maximum of the ionization-level bandwidth at the eclipsed geometry. This is the geometry for which the overlap between the two highest-occupied π-monomeric levels is optimized.

To summarize the CNDO/2 results, one might say that they lead to nothing startlingly new or different from the earlier extended-Hückel results. This arises essentially from the close cancellation between the core-core and electronic contributions to the energy as the dimer geometry is varied. The CNDO/2 results, therefore, indicate that mean field-electrostatic effects including exchange are relatively unimportant in determining the stacking geometry. The somewhat deeper well depths associated with the TCNQ eclipsed geometries lead one to suspect the ability of zero differential overlap schemes to predict the relative well depths associated with two or more energy minima. Both molecular-orbital programs also show that the banding of the ionization and the affinity levels of the respective donor and acceptor stacks contributes significantly to the total energy. The difference between the number of total energy minima for TCNQ and for TTF has also been shown to be related simply to the symmetry and nodal structure of these monomeric π orbitals. The inability of these molecular-orbital programs to predict the observed intermolecular separation between molecules on a segregated stack seems to imply, however, that the story is far from being qualitatively complete.

5.4 Interactions Between Closed-Shell Neutral TTF Molecules: Hard Sphere Packing and Atom-Atom Potentials in Crystalline TTF

A comparison of the stacking of molecules in neutral crystalline TTF and in TTF-TCNQ provides a useful clue concerning the origin of the stacking geometry in these materials. Figure 5.11 is a view perpendicular to the molecular planes of the TTF molecules on a stack in neutral crystalline TTF (Fig.5.11a) and in TTF-TCNQ (Fig. 5.11b). Projections of two neighboring molecules are shown. The intermolecular separation in neutral TTF is 3.62 Å, whereas the separation between TTF molecules in TTF-TCNQ is 3.47 Å. It is interesting to note that aside from the small relative shift between molecules along the short molecular axis in neutral TTF, the slip along the long molecular axes is similar in both materials. Since the interactions

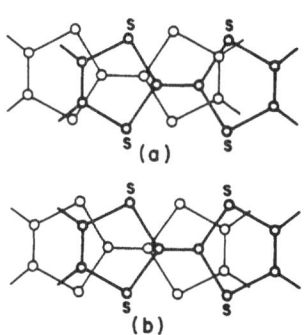

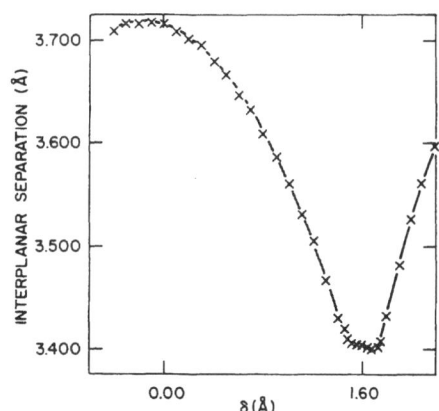

Fig. 5.11. TTF stacking in (a) crystal-line TTF [5.8], (b) TTF-TCNQ [5.9]

Fig. 5.12. Intermolecular separation of the neutral TTF dimer at closest approach as a function of relative molecular displacement

between the molecules in neutral TTF are between closed-shell molecules, this suggests that such interactions might stabilize this stacking geometry in TTF-TCNQ as well as in neutral TTF.

In general, two empirical procedures have had success in predicting the structure of molecular crystals [5.4]. First, there have been numerous investigations of possible close-packed structures composed of molecules consisting of hard-sphere atoms with van der Waals radii. There is a general tendency of molecular crystals to adopt these close-packed structures. Also, atom-atom potentials have been used to search for energy minima associated with stable structures in molecular crystals. We will use these two procedures to investigate possible stable conformations of a TTF dimer composed of two neutral monomers.

Figure 5.12 is a plot of the possible closest approach between the molecular planes of TTF when the two monomers composing the dimer are slipped over each other. Appropriate spheres having van der Waals radii have been circumscribed about the atoms [5.30]. The distance of closest approach was found by calculating the separations between atoms on the different molecules and determining when any such separation was equal to the sum of the two relevant van der Waals radii. The distance of closest approach is plotted as a function of slip, δ. The value of slip associated with the eclipsed dimer geometry is now $\delta = 0.0$. One sees that there is a distance of closest approach at the value of slip $\delta = 1.6$ observed in neutral TTF. As one slips the molecules away from this distance, towards the eclipsed conformation, the large sulfur radii come into play and move the molecules apart. This repulsion is due essentially to the increase in the overlapping of sulfur orbitals and hence to an increase in kinetic energy. The distance of closest approach at the eclipsed geometry is the largest such distance over the range of slip investigated. Hence, the interaction between closed-shell molecules favors the slipped conformation. The

122

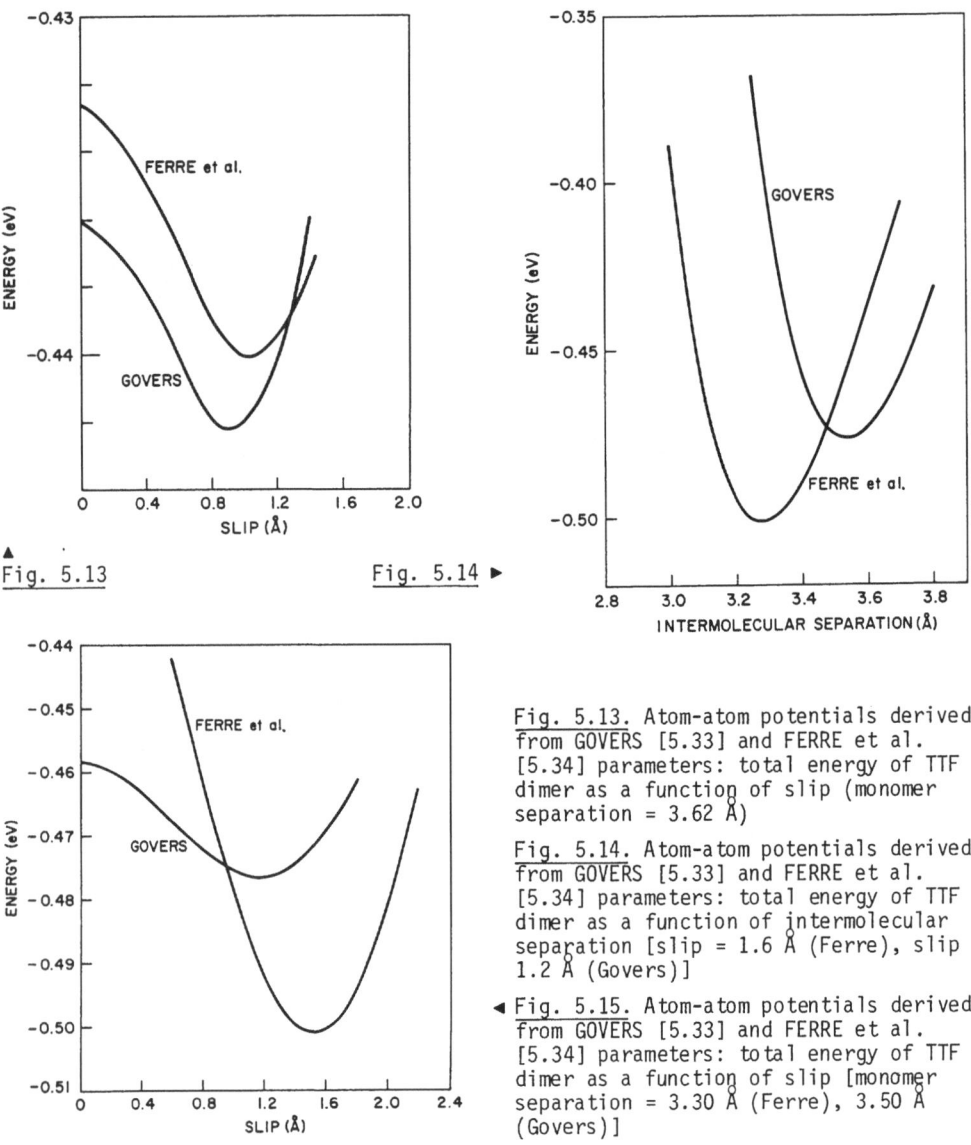

Fig. 5.13

Fig. 5.14 ▶

Fig. 5.13. Atom-atom potentials derived from GOVERS [5.33] and FERRE et al. [5.34] parameters: total energy of TTF dimer as a function of slip (monomer separation = 3.62 Å)

Fig. 5.14. Atom-atom potentials derived from GOVERS [5.33] and FERRE et al. [5.34] parameters: total energy of TTF dimer as a function of intermolecular separation [slip = 1.6 Å (Ferre), slip = 1.2 Å (Govers)]

◀ Fig. 5.15. Atom-atom potentials derived from GOVERS [5.33] and FERRE et al. [5.34] parameters: total energy of TTF dimer as a function of slip [monomer separation = 3.30 Å (Ferre), 3.50 Å (Govers)]

molecular-orbital results, on the other hand, show that the emptying of the ioniza-tion level of TTF leads to a stabilization of the eclipsed geometry. As a result one expects that a competition exists between the two modes of stacking. The mode of stacking then should be determined by the predominance of either the closed-shell interaction E_N or the change in interaction energy ΔE_{CT} as a result of charge trans-fer. This is consistent with LA PLACA's observation that among the small number of TTF compounds known, the larger the value of the charge transfer, the more likely is an eclipsed stacking geometry. Similar hard-sphere calculations performed on the neutral TCNQ dimer indicate that neutral TCNQ should have less of a preference for a particular stacking geometry than TTF. As one slips the TCNQ monomers over each other, the modulation in distance of closest approach between the molecules is significantly less than found for the TTF dimers. It is of interest that for the neutral TCNQ structure investigated [5.31], the molecules do not form clearly identifiable segregated stacks. One can, therefore, conclude that in organic salts, TCNQ forms either eclipsed or slipped segregated stacks as a result of the filling of its affinity level, while the closed-shell interactions in TTF can also play an important role in determining the stacking geometry in the salts of this organic donor molecule.

Atom-atom potentials have been recently used to determine the cohesive energy of TTF and TTF-TCNQ [5.32,33]. Such potentials can also be used to investigate the total energy as a function of relative orientation of the monomers of the TTF dimer. A commonly used form for the total energy of a neutral molecular crystal is (Chap.2 and [5.4])

$$E = \sum_{i,j} \left[B\, e^{-Cr_{ij}} - \frac{A}{r_{ij}^6} \right] \, . \tag{5.2}$$

The sums are over all pairs of atoms residing on different molecules. Figure 5.13 shows the variation in total energy as the monomers are slipped with respect to each other at the fixed intermolecular separation of 3.62 Å. Since the results of using atom-atom potentials are somewhat sensitive to the choice of parameters, we have used two sets of different parameters for each calculation. Since FERRE et al. [5.34] give all the necessary parameters except those for the sulfur-sulfur interaction, we use GOVERS' [5.33] sulfur-sulfur parameters along with the FERRE parameters for the first set. GOVERS' parameters are used for the second set. One sees from Fig.5.13 that the energy does go through a minimum at a value away from the eclipsed geometry. On the other hand, the minimum is shallow and somewhere between the eclipsed geometry and the slipped geometry observed in neutral TTF. The calculated result can be improved by determining the intermolecular separation for which the energy as a function of this separation is at a minimum. Figure 5.14 is a plot of total energy as a function of intermolecular separation. Both curves have been calculated for the values of relative monomer slip given in Fig.5.14. One sees that there is a minimum in the total energy in the vicinity of the observed intermolecular separation 3.62 Å in neutral TTF. In Fig.5.15, we show the total energy as a function of slip calculated with the value of intermolecular separation at the minimum shown in Fig.5.14. The difference between the energy at the eclipsed geometry and at the minimum for GOVERS' parameters is about 0.02 eV. Since every monomer interacts with two of its neighbors on the stack, this difference in energy is about 0.04 eV. The total interaction energy at the minimum is about 0.9 eV. Comparison of this value with the total calculated binding energy per monomer [5.32,33] for neutral TTF shows that most of the binding energy arises from the interactions between nearest neighbor molecules on a stack. Neither of the two sets of atom-atom parameters predict the observed molecular separation (center to center) along the stack to better than a few tenths of an angstrom. However, these results are in qualitative agreement with the observed crystalline structure. In contrast, the molecular-orbital calculations did not predict stability for any slipped conformation and did not yield an energy minimum for any reasonable monomer separation. The van der Waals interaction energy present in the atom-atom calculation is neglected in the molecular-orbital calculation. In the next section, we will show that Coulomb correlation, which has been viewed [5.35] as a "van der Waals interaction in the region of significant molecular-orbital overlap", provides an important part of the bonding of the π complex we are examining.

5.5 Density-Functional Calculation: Neutral TTF Dimer

Until recently, there were no nonempirical procedures that could treat the interaction between large closed-shell molecules accurately. The interactions between closed-shell molecules in crystals were treated by the empirical procedures discussed in the preceding section, namely, the hard-sphere and atom-atom approximations. An extremely interesting and valuable book detailing the power and general applicability of these methods is the work by KITAIGORODSKY [5.4]. In Chap.2 WILLIAMS also discusses the use of the atom-atom potential method. Even though these methods

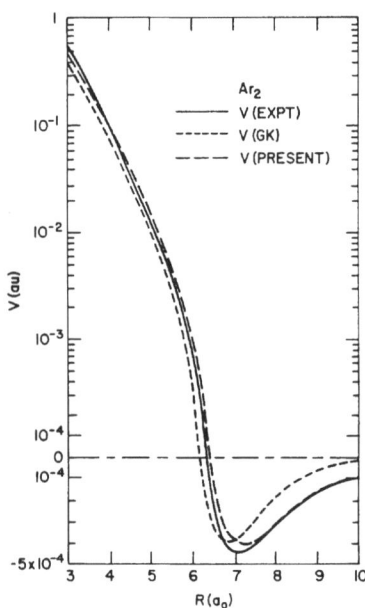

Fig. 5.16. The Ar_2 interatomic potential energy. V(EXPT): experimental values. V(GK): Gordon-Kim calculation [5.36]. V(PRESENT): Gordon-Kim calculation modified by appending to the interaction energy the long-range part of the van der Waals energy. From [5.35]

provide useful predictions concerning the expected equilibrium structures of crystals composed of neutral molecules, one is motivated to pursue the answers to questions concerning the origin of such structures with the use of a more fundamental,
nonempirical procedure. Several years ago GORDON and KIM [5.36] proposed a simplified, approximate version of local-density-functional theory to be used for calculating the interaction energy between closed-shell atoms or molecules. Some idea of
the success achieved by this procedure is demonstrated in Fig.5.16. Illustrated is
a plot of the interaction energy of two argon atoms as a function of their interatomic separation. An important thing to notice is that both calculations yield an
interaction energy minimum within 5% of the measured interatomic separation. Subsequent to these early calculation, numerous Gordon-Kim type calculations have been
performed. It is now generally agreed that these calculations lead to an accurate
location of energy minima. A recent review [5.37] of the Gordon-Kim procedure has
appeared which lists many of the relevant references. We will use this procedure to
investigate the interaction energy between the two neutral TTF monomers that compose
our dimer [5.38].

First, let us briefly discuss the ingredients of the calculation [5.38]. One first
obtains the best wave functions available for the individual monomers, in our case
the TTF molecules.

One further assumes that the total electronic density at each point in space is
just the sum of the monomer densities. In other words, the monomer wave functions
remain undistorted even when they overlap. This was originally argued to be a reasonable assumption for the rare gas atoms. It is certainly less reasonable in connection with TTF. We will, however, adopt the assumption at present.

The kinetic, exchange, and correlation interaction energies are next evaluated
with use of the following energy-density expressions obtained from uniform electron-
gas theory [5.38]:

$$E_{KE} = (3/10)(3\pi^2)^{2/3} \rho^{2/3} \tag{5.3}$$

$$E_{EX} = -(3/4)(3/\pi)^{1/3} \rho^{1/3} \tag{5.4}$$

$$E_{CORR} = -0.06156 + 0.01898 \ln(r_s) \quad , \tag{5.5}$$

where ρ is the electron density and r_s the average spacing between electrons. The
interaction energy between molecules 1 and 2 can be written as

$$E_{INT} = \int d\bar{r} [(\rho_1 + \rho_2) E_d (\rho_1 + \rho_2) - \rho_1 E_d (\rho_1) - \rho_2 E_d (\rho_2)] \quad , \tag{5.6}$$

with

$$E_d = E_{KE} + E_{EX} + E_{CORR} \quad . \tag{5.7}$$

To these three contributions to the interaction energy one adds the direct Coulomb interaction between the two monomers, i.e., the electrostatic interactions between the charge distribution of one of the monomers, electrons and cores, with the electrons and cores of the other monomer.

Since the TTF molecule is sufficiently large that Hartree-Fock quality wave functions of the type used for treating the rare gas interactions are not easily obtained, we will, for convenience, use the wave functions previously obtained from the semi-empirical molecular-orbital calculations. We have also obtained the Gaussian 70 wave function [5.39] for the TTF monomer. The most complete calculations have been performed with the Gaussian 70 (STO-3G) wave function. On the other hand, the results that will be presented are qualitatively independent of the starting wave functions.

Calculations of the interaction energy arising from the electron gas contribution, (5.6), were performed with extended-Hückel wave functions (with and without d functions on the sulfurs), a CNDO/2 wave function and a Gaussian 70 wave function. Equation (5.6) was evaluated by replacing the integral by a sum over a three-dimensional rectangular grid. A typical grid mesh is shown in Fig.5.17. The electron densities ρ, ρ_1, and ρ_2 were evaluated at each grid point. These values along with (5.3-7) enable one to calculate the interaction energy contribution from the three electron-gas terms. The grid spacing was varied to check that the result was independent of the grid interval chosen. Since the Coulomb contribution to the interaction energy is the most difficult to calculate and the location of the energy minimum does not depend on this contribution, we will defer the discussion of this contribution to later. Figure 5.18 is a plot of the interaction energy as a function of slip for the four different wave functions used to calculate the electronic density of the monomers. As we have previously stated, all of the curves exhibit the same qualitative behavior, namely, they all go through a minimum near the value of slip observed in the crystal, $\delta = 1.6$ Å. The minimum is not particularly deep; however, it should be remembered that all values given for the dimer should be multiplied by a factor of two for molecules in the crystal. In Fig.5.19 we show a more detailed breakdown of the Gaussian 70 results. The dotted curve is the sum of the two largest contributions, the kinetic energy and the exchange energy. These two terms largely cancel and therefore result in essentially no net bonding of the complex. The shallow minimum exhibited by the sum of these two terms is not found for all of the starting monomer wave functions. The presence or absence of such a shallow minimum could be sensitive to the exchange self-energy correction of the type proposed by RAE [5.40]. Again we emphasize that we are using the original version of the Gordon-Kim calculation with none of the proposed modifications [5.37]. It is of interest that the best agreement with experiment for the calculated intermolecular potential between two methane molecules has been obtained when no corrections are applied to the individual contributions of the total electron-gas energy [5.41]. Also, a number of the proposed modifications to the original Gordon-Kim theory have apparently yielded no better detailed agreement between experiment and theory for systems involving ions and/or

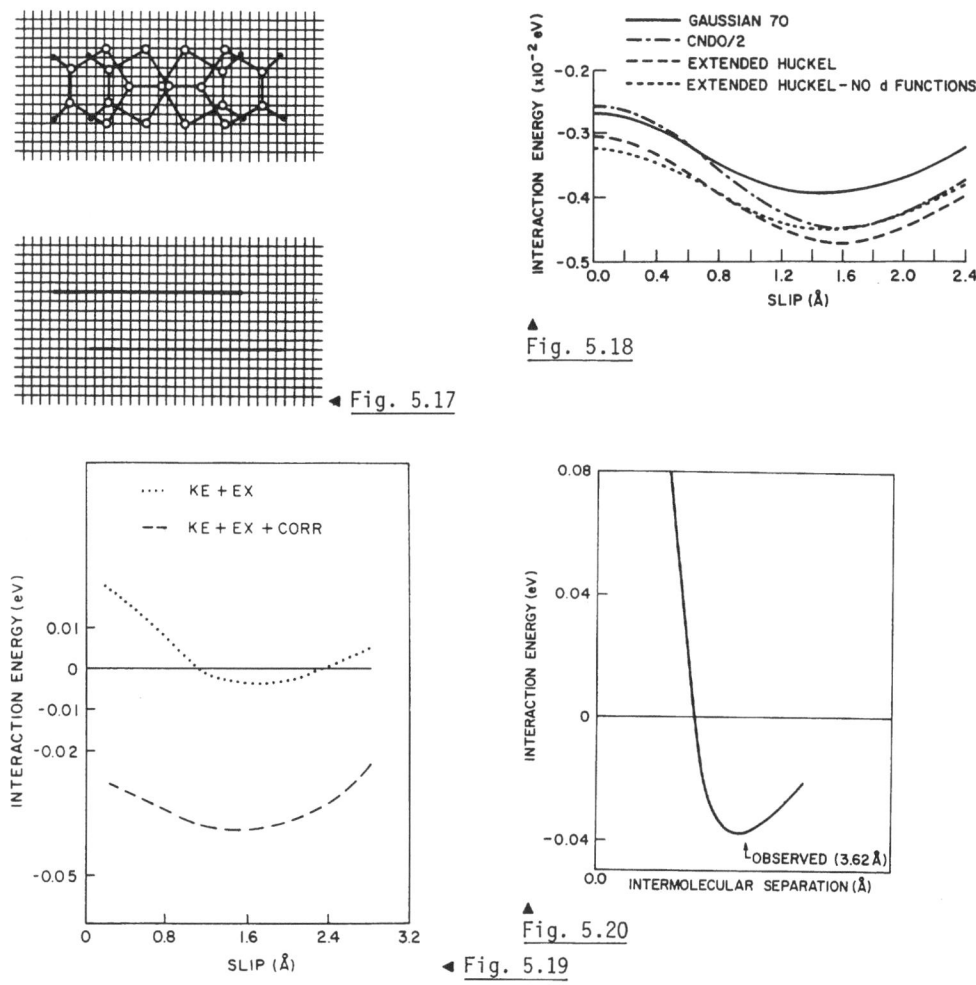

Fig. 5.17

Fig. 5.18

Fig. 5.20

◀ Fig. 5.19

Fig. 5.17. A typical grid mesh for which the Gordon-Kim type calculation is performed [5.38]

Fig. 5.18. Electron-gas contribution to the TTF interaction energy as a function of slip [5.38]

Fig. 5.19. Electron-gas contribution to the TTF interaction energy as a function of slip using Gaussian 70 wave functions showing the sum of the kinetic (KE) and exchange (EX) energies (dotted line) and the sum of the KE, EX, and correlation (CORR) energies (dashed line) [5.38]

Fig. 5.20. TTF interaction energy as a function of intermolecular separation using Gaussian 70 STO-3G wave functions in a Gordon-Kim procedure [5.38]

heavier neutral atoms [5.42]. A major contribution to the bonding of the complex is seen to be due to the correlation energy. This contribution is important in determining the location of the energy minimum. This result shows why the extended-Hückel and CNDO/2 calculations fail to yield an interaction energy minimum at negative energy as a function of slip. Since all Hartree-Fock calculations neglect electronic correlation, one suspects that any such calculation, no matter how close to the true

Hartree-Fock limit, will be incapable of yielding appropriate interaction energy minima associated with closed-shell molecules interacting at van der Waals distances. Procedures have been proposed where the long-range van der Waals interaction energy is just added to the Hartree-Fock results. As pointed out by COHEN and PACK [5.35], the correlation energy is in a sense the dispersion energy (van der Waals) in the region of the overlap.

Figure 5.20 shows the interaction energy as a function of intermolecular separation. This curve has also been obtained without the use of the Coulomb interaction. This calculation was performed for a value of slip, δ, at the minimum shown in Fig. 5.4. Again only the Gaussian 70 result is given. The interaction energy is found to go through a minimum that is near the observed separation between molecules on a stack in neutral TTF, namely near 3.62 Å. The rapid rise in interaction energy with decreasing intermolecular separation reflects the rapid increase in kinetic energy as one attempts to localize electrons by overlapping the two rigid monomer wave functions. The relatively accurate estimate by the Gordon-Kim procedure of this rapid rise in kinetic energy is one of the main reasons for the success of this method in yielding accurate interaction energy minima [5.37]. The observed intermolecular separation of 3.62 Å was the value used to obtain Fig.5.19. Calculations of the interaction energy as a function of relative slip along the shorter principal molecular axis have also been performed. The interaction energy goes through a minimum at zero for such slip, i.e., for a slip along only the major molecular axis. Therefore, the Gordon-Kim procedure with only the three electron-gas terms yields an interaction-energy minimum occurring at a relative separation in three dimensions between closed-shell TTF molecules that is about 0.1 Å from the actual observed separation in the crystal. This result is in agreement with the hard-sphere and atom-atom potential results, namely, that the neutral TTF stacking should be slipped.

The electrostatic or Coulomb interactions between the charge distributions of the neutral molecules are difficult to treat even though they are the simplest contribution to the interaction energy to visualize. These interactions consist of a core-core, core-electron, and an electron-electron part. If one strips the valence electrons off the molecules, the remaining core-core interaction energy is on the order of 10^4 eV. In the presence of the electrons, the total electrostatic interaction energy between molecules is on the order of 10^{-2} eV. The electrons, therefore, closely screen the core-core interaction and the calculations must be set up carefully with this in mind. If one is going to investigate the variation in electrostatic energy with stacking geometry, the calculations must be performed precisely to at least one part in 10^7 of the core-core repulsion. For spherically-symmetric atomic-charge distributions, a large part of this cancellation is taken care of by doing two of the three integrations analytically. We have not approximated the TTF charge distributions by spherical charge distributions. Arguing further along these lines, one expects that the monomer electronic densities obtained from a non-self-consistent wave function, e.g., an extended-Hückel wave function, would be a poor choice to use in the calculation of the Coulomb interaction term. This has been our experience. The Gaussian 70 wave function for the neutral TTF monomer was obtained primarily for use in the electrostatic part of the Gordon-Kim calculation. We found that the QCPE version [5.39] of Gaussian 70 did not converge for the TTF molecular geometry used. Convergence was achieved by averaging the density matrix over the values obtained for the two latest iterations before recalculating the Fock matrix. The neutral TTF monomer treated with an STO-3G basis requires exactly 70 Slater orbitals. Each Slater orbital is decomposed into three Gaussians. The Gaussian basis therefore consists of 210 functions. Since the Coulomb interaction energy is in principle directly integrable in a Gaussian basis, one might expect that this can be simply calculated. This is not so. Since the total number of occupied molecular orbitals is 52, the total number of calculations required would be approximately $(52)^2 (210)^4 \approx 10^{12}$. This procedure is still not feasible even if one collapses the core orbitals onto a point charge nuclear core. Other techniques have been proposed to calculate the Coulomb interaction term in the Gordon-Kim procedure. It is possible to expand the charge density directly in terms of a set of basis functions [5.37]. One can also construct spherical charge distributions centered about the respective atoms composing the molecule [5.41]. A good part of the calculation can be then done analytically. We have not adopted these techniques. We have calculated the electrostatic interactions directly in real space by calculating the electron-electron repulsion with use of two three-dimensional grids. To this was added the electron-core and core-core interaction calculated in a similar manner. Core orbital charges were used in a renormalization of the nuclear charges with use of a Mulliken population analysis. Certain general precautions were taken. Total charge neutrality of the system was maintained by slightly readjusting the core charges after the total electronic density was collected at all of the grid points. The calculations were therefore performed by fixing two different three-dimensional grids with respect to each of the monomer nuclear positions. The finest grid that we found reasonable to use involved 18081 grid points. The grid spacing was 0.5 Å and was therefore relatively coarse on a molecular scale. Since the monomer displacements investigated were comparable to the grid-point separation, "grid effects" were superimposed on the variation of electrostatic interaction energy with relative monomer displacement. Such effects were handled by recognizing that there are sets of relative monomer displacements for which the relative orientation of the two grids is identical. Such equivalent grid positions are illustrated in Fig.5.21 for a one-dimensional grid. A point designated as the origin on one of the grids may be placed at any one of the positions designated by a cross while still maintaining the same relative orientation between the two grids — the one with its origin at one of the crosses and the other designated by the filled circles. One then expects the calculated Coulomb interaction energy to vary smoothly, with no grid effects, when calculated with the two grids at these equivalent grid positions. In this way, families of smooth curves are generated, each associated with a different set of equivalent grid positions. Since grid

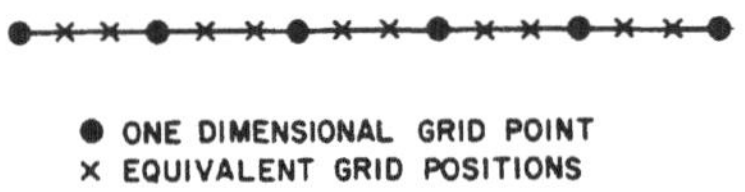

130

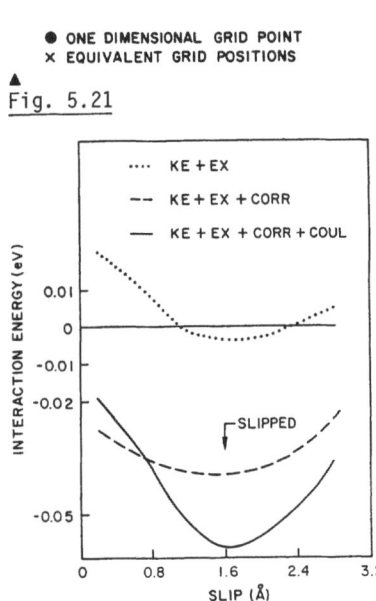

● ONE DIMENSIONAL GRID POINT
× EQUIVALENT GRID POSITIONS

▲
Fig. 5.21

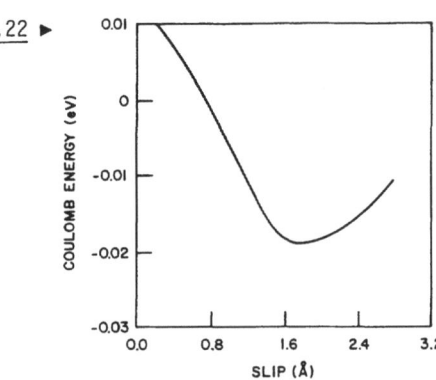

Fig. 5.22 ▶

Fig. 5.21. Equivalent grid positions on a one-dimensional grid

Fig. 5.22. TTF Coulomb interaction energy as a function of slip using the Gordon-Kim procedure [5.38]

◀ Fig. 5.23. TTF total interaction energy as a function of slip showing various combinations of kinetic (KE), exchange (EX), correlation (CORR), and Coulombic (COUL) energies

effects contribute to spurious electron-electron repulsions, all curves are then translated to the values obtained when the grids are farthest apart. This is how Fig.5.22 has been obtained.

From Fig.5.22, one sees that as the molecules are slipped away from the eclipsed stacking geometry, the electrostatic interaction energy changes from being repulsive to being attractive in the vicinity of the slipped geometry. In order to understand this behavior qualitatively, it is useful to separate two different features associated with the monomer electronic charge density. The charge transfer between the various atomic constituents produces quadrupolar and higher-order moments. When the monomers are in the eclipsed geometry, then like atomic charges are head on, and this contributes to the repulsion between molecules. As the molecules are slipped with respect to each other, this repulsion decreases, and the electrostatic interaction becomes bonding near the observed slipped geometry. The bonding contribution is a result of electronic delocalization away from the nuclear cores. Remember that for two interacting argon atoms, if one collapses the electronic distribution onto the nuclei, the electrostatic interaction energy is exactly zero. It is of interest that if the TTF electron densities are collapsed onto the cores, the electrostatic interaction energy is repulsive for all values of slip investigated. This procedure of assigning net charge densities to point charges at the nuclear positions is a procedure that has been generally adopted in calculations of the Madelung energies of neutral molecular crystals [5.32,33] and organic salts (Chap.4 and [5.43]). The significance of such numerical values will therefore be subject to question until electronic delocalization is also taken into account in such calculations. Figure 5.23

shows the total interaction energy between monomers as a function of slip. The
Coulomb interaction energy has been added to the results shown in Fig.5.19. The
effect of adding this energy is, therefore, to sharpen the total interaction energy
minimum and make it somewhat deeper. Whereas the position of the minimum is near
the position found in neutral TTF, the depth of the minimum is almost an order of
magnitude smaller than the result obtained from atom-atom potentials [5.32,33]. We
had previously pointed out that the results of the atom-atom potentials are consistent with the measured [5.32] binding energy per molecule for neutral TTF. The calculated Gordon-Kim well depth is, therefore, anomalously low. One expects this to
be mainly due to the neglect of the long range part of the dispersion (van der Waals)
interaction. π-complexes such as the one we are presently investigating involve a
relatively large number of pairs of atoms interacting at distances beyond the region of overlap where electronic correlation reasonably describes dispersion. A simple estimate of the long range van der Waals part indicates that it could make up
the discrepancy between the calculated Gordon-Kim well depth and the measured binding energy of neutral TTF. The van der Waals constants, e.g., C_6, C_8, and C_{10} are,
however, required between the relevant atom pairs before the long range van der Waals
interaction energy can be calculated with any accuracy [5.35].

We should emphasize that the location of the energy minimum obtained with only
the three electron-gas contributions of the Gordon-Kim model is near the position
observed in neutral TTF. Many modifications of the Gordon-Kim procedure have been
proposed [5.37]. Two of the more important ones we have already mentioned, namely,
appending the long-range van der Waals interaction energy and correcting the spurious self-exchange contribution. For simplicity, we have not adopted any of these
modifications. Our objective has been somewhat different from the objective of previous Gordon-Kim calculations. We have not been interested in comparing the dependence of the interaction energy as a function of intermolecular separation with the
results of experiment or the calculated zero-temperature value of compressibility
with measurement. These experimental values are not presently available. We were
essentially looking on a much cruder scale for the presence or absence of any indication of stability for the slipped stacking geometry observed in neutral crystalline TTF. Such indication is provided by calculating only the electron-gas contributions with the original Gordon-Kim procedure. The success of this simple procedure
in yielding the intermolecular separation suggests that such a procedure might be
valuable in resolving similar questions involving interactions between even làrger
closed-shell molecules. If one is willing to neglect calculation of the electrostatic term, the calculation can be scaled up significantly. If the monomer units
do however have a dipole moment, neglect of the electrostatic term may be unreason-
able since dipole interactions could contribute significantly to the stabilization
of a particular equilibrium geometry.

It is not quite clear how accurate the shape or well depth of the interaction
energy will be, even after the suitable modifications have been made to the original

Gordon-Kim procedure. One fundamental assumption of the theory is that the monomer wave functions remain undistorted even when they overlap significantly. This is certainly less valid for the present case than for interacting rare-gas atoms. The ionization energy of an electron in the highest occupied molecular-orbital of TTF is roughly half the value of the rare-gas ionization energies. It is also commonly stated that molecules such as TTF and TCNQ are highly polarizable. Any polarization energy contribution in the neutral structure is not taken into account by the Gordon-Kim calculation on the neutral dimer. These as well as other questions remain. On the other hand, our results suggest that gross conformational features such as interaction energy minima may be generally determined with accuracy by the Gordon-Kim procedure.

5.6 Density-Functional Calculation: Open Shell TTF Dimer

Since the calculations discussed in the previous sections on the interactions between neutral molecules were originally motivated by questions concerning stacking in TTF-TCNQ, we should return to this question. Let us attempt to modify the molecular-orbital calculations of the type discussed in Sect.5.3 so that they treat the closed-shell part of the interaction E_N more accurately. Remember, E_N is the interaction energy between neutral molecules constrained to adopt the organic-salt structure. Let us assume that the total energy E_T^+ of the TTF dimer with some degree of ionization has been obtained from some molecular-orbital program. We would then write

$$E_T^+ = E_N + \Delta E_{CT}^+ \quad . \tag{5.8}$$

E_N would then be defined as the result of running the molecular-orbital program at the same geometry for the neutral dimer. Since we know E_N does not characterize the interactions between closed-shell molecules accurately for reasons we have stated, we will subtract this from the total energy E_T^+ and add the Gordon-Kim expression for the interaction energy between neutral monomers E_N^{GK}, thus obtaining a new energy $E_T^{'+}$

$$E_T^{'+} = E_T^+ - E_N + E_N^{GK} = E_N^{GK} + \Delta E_{CT}^+ \quad . \tag{5.9}$$

Note that since we have subtracted a total dimer energy and added an interaction energy between monomers, the new energy $E_T^{'+}$ is neither a total energy nor an interaction energy. This does not matter, however, since we are only interested in the variation of $E_T^{'+}$ with changes in the relative spatial orientation of the two monomers while keeping the monomer geometry fixed. The energy from (5.9) at a fixed relative monomer geometry then consists of two parts. The first part, E_N^{GK}, is the

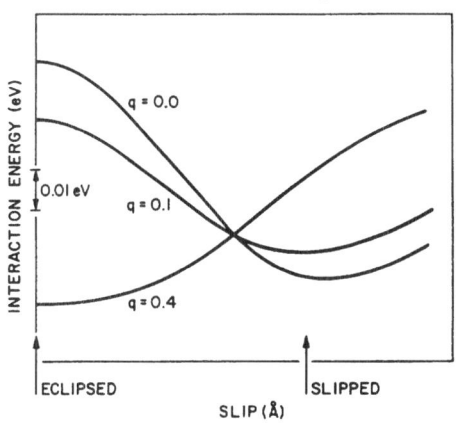

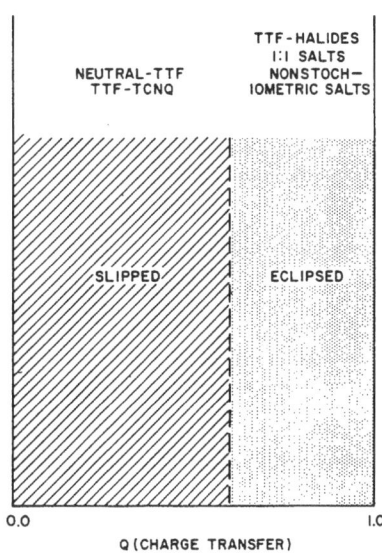

Fig. 5.24. Interaction energy $E_T^{'+}$ of TTF dimer as a function of slip: parametrized with respect to charge transfer, q

Fig. 5.25. Phase diagram for stacking geometry: dependence of structure on charge transfer

Gordon-Kim interaction energy between neutral molecules. The second part is the change in dimer energy as a result of ionization. For an independent electron molecular-orbital calculation such as any Hückel scheme, this change in energy is just the ionization energy of the uppermost filled orbital of the neutral dimer multiplied by the fractional ionization of this level. In a self-consistent molecular-orbital calculation, such as CNDO/2, ΔE_{CT}^+ cannot be simply obtained from the single-particle levels of the neutral dimer. For this case, as previously stated, ΔE_{CT}^+ is obtained as the difference between the total energy of the fractionally ionized dimer and the neutral dimer. ΔE_{CT}^+ calculated in this manner does not include the effect of electronic correlation and is subject to the approximations of the particular molecular-orbital program used. On the other hand, ΔE_{CT}^+ does qualitatively include the major contributions that arise from covalency as a consequence of emptying the ionization level of the dimer.

As we have shown, E_N^{GK} goes through an energy minimum in the vicinity of the slipped geometry. In Sect.5.3 we had shown that the charge transfer part, ΔE_{CT}^+, went through a minimum at the eclipsed geometry. These two contributions, therefore, compete with each other in establishing the equilibrium geometry. To illustrate this, we have calculated the energy $E_T^{'+}$ as a function of slip along the long molecular axis. E_N^{GK} was just chosen from the values calculated in the previous section. ΔE_{CT}^+ was calculated with the extended-Hückel procedure. The results are only of heuristic value due to the approximate nature of the calculations. For simplicity we have just fixed the equilibrium separation appropriate to the neutral structure. The results presented in Fig.5.24 therefore indicate only qualitatively what can be expected as one

varies the charge transfer. In the figure, energy is plotted as a function of slip
for three different values of charge transfer. One sees that for small values of
charge transfer or ionization of the TTF dimer the equilibrium geometry is slipped.
For values of charge transfer greater than some critical value the stable dimer
geometry will be eclipsed. This suggests the following phase diagram shown in Fig.
5.25 for TTF stacking. Below a certain value of charge transfer, the stacking geometry will be slipped as observed in neutral TTF and TTF-TCNQ. The nonstoichiometric
halide salts of TTF with segregated stacks as well as the 1:1 salts that consist of
dimers exhibit eclipsed stacking [5.10,11]. It should be pointed out that an exact
one-to-one correspondence between charge transfer and stacking geometry is however
violated by the disordered halides. Disordered TTF-Br with an average charge transfer of approximately 0.55 is eclipsed, whereas TTF-TCNQ with a charge transfer of
0.59 is slipped. Interactions between donor and acceptor molecules also contribute
to the determination of the stacking geometry and a more complete treatment should
include these as well. For TTF-TCNQ interactions between molecules along the a_ direction may contribute to the stabilization of the slipped stack geometries through the
interactions between the sulfur and nitrogen atoms. In general, the extrapolation
of the TTF dimer results to organic salts should be of greater validity the more
isotropic are the interactions that involve the anion.

5.7 Conclusions

The work presented in this chapter has focused on the interactions between two TTF
monomers at or near van der Waals separations. To a lesser degree, interactions between TCNQ monomers were also examined. The motivation was twofold. First, it was
of interest to understand why molecular-orbital programs have yielded anomalous results concerning equilibrium geometries. Second, it was of interest to see if dimer
calculations could tell us anything useful about the stacking observed in the neutral
crystals or the organic salts. Calculations performed on the neutral dimer indicate
that the electronic correlation energy (van der Waals energy) is important in bonding the neutral complex. This is, of course, what one would have expected. The important point, however, is that in an organic salt where covalence effects are important, these effects are sufficiently weak that electronic correlation can still
play an important role in bonding the complex and in determining a particular stable
geometry. It is, therefore, inappropriate to calculate total dimer energies with
some Hartree-Fock procedures and expect generally meaningful results. This might
happen in the event that the covalency interactions alone are responsible for some
particular geometrical feature, e.g., the eclipsed as well as slipped stacking geometries of TCNQ molecules. On the other hand, the intermolecular separation of TCNQ
monomers at these geometries is not predicted by the molecular-orbital programs.

The results we have obtained for the TTF dimer indicate that below a certain degree of ionization the dimer is stable at the slipped geometry. This is consistent with the geometry observed in certain organic salts containing this donor molecule. But since the energy difference between the eclipsed and slipped geometries is small (≈ 0.05 eV), one can expect covalency effects, as well as other intermolecular interactions, to make important contributions to the equilibrium stacking geometry adopted by the salt. Consequently, an unambiguous quantitative determination of the segregated stacking geometry in an organic salt is not a simple task. It should at least be recognized, however, that this problem is as close to the problem of determining the structure of neutral molecular crystals as it is to the problem of determining the structure of a strongly covalently bound solid.

Acknowledgements. I would like to thank Evelyn Marino for assisting in the preparation of the manuscript.

References

5.1 J.S. Miller, A.J. Epstein (eds.): *Synthesis and Properties of Low-Dimensional Materials*, Proc. of the New York Academy Conf., 1977, Ann. N.Y. Acad. Sci. *313* (1978)
5.2 W.E. Hatfield (ed.): *Molecular Metals*, NATO Conf. Proc., Les Arcs, France 1978 (Plenum, New York 1978)
5.3 J.T. Devreese, R.P. Evrard, V.E. Van Doren (eds.): *Highly Conducting One-Dimensional Solids* (Plenum, New York 1979)
5.4 A.I. Kitaigorodsky: *Molecular Crystals and Molecules* (Academic, New York 1973)
5.5 F.H. Herbstein: In *Perspectives in Structural Chemistry*, Vol.4, ed. by J.D. Dunitz, J.A. Ibers (John Wiley, New York 1971) p.166
5.6 R.P. Shibaeva, L.O. Atomvyan: Sov. J. Struct. Chem. *13*, 514 (1972)
5.7 I.F. Shchegolev: Phys. Status Solidi A *12*, 9 (1972)
5.8 T.J. Kistenmacher, T.E. Phillips, D.O. Cowan: Acta Crystallogr. *B30*, 763 (1974)
5.9 T.E. Phillips, T.J. Kistenmacher, J.P. Ferraris, D.O. Cowan: J. Chem. Soc. Chem. Comm. *471* (1973)
5.10 S.J. LaPlaca, P.W.R. Corfield, R. Thomas, B.A. Scott: Sol. St. Comm. *17*, 635 (1975)
5.11 B.A. Scott, S.J. LaPlaca, J.B. Torrance, B.D. Silverman, B. Welber: J. Am. Chem. Soc. *99*, 6631 (1977)
5.12 M. Konno, T. Ishii, Y. Saito: Acta Crystallogr. *B33*, 763 (1977)
5.13 M. Konno, Y. Saito: Acta Crystallogr. *B31*, 2007 (1975)
5.14 A.J. Berlinsky: Contemp. Phys. *17*, 331 (1976)
5.15 D.B. Chesnut, R.W. Moseley: Theor. Chim. Acta *13*, 230 (1969)
5.16 E.H. Lieb: Rev. Mod. Phys. *48*, 553 (1976)
5.17 L.B. Coleman, M.J. Cohen, D.J. Sandman, F.G. Yamagishi, A.F. Garito, A.J. Heeger: Solid State Commun. *12*, 1125 (1973)
5.18 A.J. Berlinsky, J.F. Carolan, L. Weiler: Solid State Commun. *19*, 1165 (1976)
5.19 R. Hoffmann: J. Chem. Phys. *39*, 1397 (1963); Quantum Chemistry Program Exchange, Program 344 (Department of Chemistry, Indiana University, Bloomingotn, IN 47401)
5.20 J.P. Lowe: J. Am. Chem. Soc. *102*, 1262 (1980)
5.21 J.A. Pople, D.L. Beveridge: *Approximate Molecular Orbital Theory* (McGraw-Hill, New York 1970); Quantum Chemistry Program Exchange, Program 141 (Dept. of Chemistry, Indiana University, IN 47401)
5.22 M.J.S. Dewar: *The Molecular Orbital Theory of Organic Chemistry* (McGraw-Hill, New York 1969)

5.23 B.D. Silverman: J. Chem. Phys. *70*, 1614 (1979)
5.24 A.R. Gregory: In *The Jerusalem Symposia on Quantum Chemistry and Biochemistry*, Vol.VI, ed. by E.E. Bergmann, B. Pullman (Academic, New York 1974) p.23
5.25 J.B. Torrance, J.J. Mayerle, K. Bechgaard, B.D. Silverman, Y. Tomkiewicz: Phys. Rev. *B22*, 4960 (1980)
5.26 A.N. Bloch, D.O. Cowan, K. Bechgaard, R.E. Pyle, R.H. Banks: Phys. Rev. Lett. *34*, 1561 (1975)
5.27 I. Shirotani, H. Kobayashi: Bull. Chem. Soc. Jpn. *46*, 2595 (1973)
5.28 D.R. Salahub, R.P. Messmer, F. Herman: Phys. Rev. *B13*, 4252 (1976)
5.29 S.J. LaPlaca: Lecture presented at the New York Academy of Sciences Conference on the Synthesis and Properties of Low Dimensional Materials, June 13-16, 1977, New York, NY
5.30 B.D. Silverman: J. Chem. Phys. *71*, 3592 (1979)
5.31 R.E. Long, R.A. Sparks, K.N. Trueblood: Acta Crystallogr. *18*, 932 (1965)
5.32 D.J. Sandman, A.J. Epstein, J.S. Chickos, J. Ketchum, J.S. Fu, H.A. Scheraga: J. Chem. Phys. *70*, 305 (1979)
5.33 H.A. Govers: Acta Crystallogr. *A34*, 960 (1978)
5.34 Y. Ferre, E.J. Vincent, J. Metzger, A. Samat, R. Guglielmetti: Tetrahedron *30*, 787 (1974)
5.35 J.S. Cohen, R.T. Pack: J. Chem. Phys. *61*, 2372 (1974)
5.36 R.G. Gordon, Y.S. Kim: J. Chem. Phys. *56*, 3122 (1972)
5.37 M.J. Clugston: Adv. Physics *27*, 893 (1978)
5.38 B.D. Silverman: J. Chem. Phys. *72*, 5501 (1980)
5.39 W.J. Hehre, R.F. Stewart, J.A. Pople: J. Chem. Phys. *51*, 2657 (1969); Quantum Chemistry Program Exchange, Program 236 (Dept. of Chemistry, Indiana University, Bloomington, IN 47401)
5.40 A.I.M. Rae: Chem. Phys. Lett. *18*, 574 (1973)
5.41 G.C. Tabisz: Chem. Phys. Lett. *52*, 125 (1977)
5.42 M. Waldman, R.G. Gordon: J. Chem. Phys. *71*, 1325 (1979)
5.43 B.D. Silverman, S.J. LaPlaca: J. Chem. Phys. *69*, 2585 (1978)

Subject Index

---

*when specified after a page number, D = Dimer energy, L = Lattice energy, P = Pair potential energy

O. Madelung

Introduction to Solid-State Theory

Translated from the German by B. C. Taylor
1978. 144 figures. XI, 486 pages
(Springer Series in Solid-State Sciences, Volume 2)
ISBN-13: 978-3-642-81579-9

Contents:
Fundamentals. – The One-Electron Approximation. – Elementary Excitations. – Electron-Phonon Interaction: Transport Phenomena. – Electron-Electron Interaction by Exchange of Virtual Phonons: Superconductivity. – Interaction with Photons: Optics. – Phonon-Phonon Interaction: Thermal Properties. – Local Description of Solid-State Properties. – Localized States. – Disorder. – Appendix: The Occupation Number Representation.

E. A. Silinsh

Organic Molecular Crystals

Their Electronic States

Translated from the Russian by J. Eiduss in collaboration with the author
1980. 135 figures, 54 tables. XVII, 389 pages
(Springer Series in Solid-State Sciences, Volume 16)
ISBN-13: 978-3-642-81579-9

Contents:
Introduction: Characteristic Features of Organic Molecular Crystals. – Electronic States of an Ideal Molecular Crystal. – Role of Structural Defects in the Formation of Local Electronic States in Molecular Crystals. – Local Trapping Centers for Excitons in Molecular Crystals. – Local Trapping States for Charge Carriers in Molecular Crystals. – Summing Up and Looking Ahead. – References. – Additional References with Titles. – Subject Index.

A. I. Kitaigorodsky

Mixed Crystals

1981.
(Springer Series in Solid-State Sciences, Volume 33)
ISBN-13: 978-3-642-81579-9

Mixed Crystals presents a practical analysis of the structural crystallography of mixed systems. The author demonstrates that all mixed systems - be they metallic, inorganic, organic or polymeric in nature – exhibit the tendency toward the closest possible packing of particles, obeying the close packing principle regardless of the intermolecular forces involed. He shows how the investigative methods and results developed for organic substances can be transferred to other classes of substances as well.

The approach described in this book will prove useful to chemists, engineers, physicists, biologists, and crystallographers in predicting the structure of many kinds of mixed crystals and in engineering various materials with particular properties.

This book is devoted to the atomic, molecular, cluster and microcrystalline structure of mixed crystals, covering all classes of substances – organic, inorganic, metallic and polymeric. The author demonstrates that the structure of these systems may be interpreted and predicted in terms of the close-packing principle.

Springer-Verlag
Berlin
Heidelberg
NewYork

Y. N. Molin, K. M. Salikhov, K. I. Zamaraev

Spin Exchange

Principles and Applications in Chemistry and Biology

1980. 68 figures, 41 tables. XI, 242 pages
(Springer Series in Chemical Physics, Volume 8)
ISBN-13: 978-3-642-81579-9

Contents:
Introduction. – Theory of Spin Exchange. – Experimental
Measurement of Spin Exchange Rate. – Spin Exchange in
Chemistry and Biology. – References. – Subject Index.

Theory of Chemisorption

Editor: J. R. Smith
1980. 116 figures, 8 tables. XI, 240 pages
(Topics in Current Physics, Volume 19)
ISBN-13: 978-3-642-81579-9

Contents:
J. R. Smith: Introduction. – *S. C. Ying:* Density Functional
Theory of Chemisorption on Simple Metals. – *J. A. Appel-
baum, D. R. Hamann:* Chemisorption on Semiconductor Sur-
faces. – *F. J. Arlinghaus, J. G. Gay, J. R. Smith:* Chemisorption
on d-Band Metals. – *B. Kunz:* Cluster Chemisorption. –
T. Wolfram, S. Ellialtioğlu: Concepts of Surface States and
Chemisorption on d-Band Perovskites. – *T. L. Einstein,
J. A. Hertz, J. R. Schrieffer:* Theoretical Issues in Chemi-
sorption.

M. Lannoo, J. Bourgoin

Point Defects in Semiconductors I

Theoretical Aspects

1981. 87 figures, approx. 27 tables. Approx. 300 pages
(Springer Series in Solid-State Sciences, Volume 22)
ISBN-13: 978-3-642-81579-9

Contents:
Atomic Configuration of Point Defects. – Effective Mass
Theory. – Simple Theory of Deep Levels in Semiconduc-
tors. – Many-Electron Effects and Sophisticated Theories of
Deep Levels. – Vibrational Properties and Entropy. –
Thermodynamics of Defects. – Defect Migration and Dif-
fusion. – References. – Subject Index.

Springer-Verlag
Berlin
Heidelberg
New York